AF677995

Composite Materials for Civil Engineering Applications

Composite Materials for Civil Engineering Applications

Guest Editors

Yanshuai Wang

Dong Guo

Jun He

Bai Zhang

MDPI

Basel • Beijing • Wuhan • Barcelona • Belgrade • Novi Sad • Cluj • Manchester

Guest Editors
Yanshuai Wang
College of Civil and Transportation Engineering
Shenzhen University
Shenzhen
China

Dong Guo
School of Civil Engineering and Transportation
Guangzhou University
Guangzhou
China

Jun He
School of Civil Engineering
Chang'an University
Xi'an
China

Bai Zhang
School of Civil Engineering
Changsha University of Science and Technology
Changsha
China

Editorial Office
MDPI AG
Grosspeteranlage 5
4052 Basel, Switzerland

This is a reprint of the Special Issue, published open access by the journal *Journal of Composites Science* (ISSN 2504-477X), freely accessible at: https://www.mdpi.com/journal/jcs/special_issues/P1Z0W3D2JC.

For citation purposes, cite each article independently as indicated on the article page online and as indicated below:

Lastname, A.A.; Lastname, B.B. Article Title. *Journal Name* **Year**, *Volume Number*, Page Range.

ISBN 978-3-7258-7162-9 (Hbk)
ISBN 978-3-7258-7163-6 (PDF)
https://doi.org/10.3390/books978-3-7258-7163-6

© 2026 by the authors. Articles in this reprint are Open Access and distributed under the Creative Commons Attribution (CC BY) license. The reprint as a whole is distributed by MDPI under the terms and conditions of the Creative Commons Attribution-NonCommercial-NoDerivs (CC BY-NC-ND) license (https://creativecommons.org/licenses/by-nc-nd/4.0/).

Contents

About the Editors

Yanshuai Wang

Yanshuai Wang is a full professor within the College of Civil and Transportation Engineering, Shenzhen University. Prof. Wang is a recipient of the National-Level Young Talent program and holder of both Guangdong Outstanding Talent (Category B) and Shenzhen High-Level Overseas Talent. He is the Associate Director of the Shenzhen Key Laboratory of Low-Carbon Construction Materials and Technology, and he serves on the Editorial Board of *Journal of Composites Science* (IF = 3.7) and is a Youth Editorial Board Member of *Journal of Building Materials*. He has been listed in the World's Top 2% Scientists (2023–2025).

His long-term research focuses on zero-carbon/low-carbon civil engineering materials and their applications, corrosion degradation inhibition and protection of marine concrete, and the safe disposal and utilization of solid waste in construction materials. He has led three projects supported by the National Natural Science Foundation of China, one sub-project of the National Key R&D Program, and six provincial/municipal-level research projects, including a Key Guangdong–Hong Kong Cooperation Project. His achievements include being awarded the First Prize of the Guangdong Natural Science Award, the First Prize of the Tianjin Science and Technology Progress Award, and the Second Prize of the China Communications Construction Group Science and Technology Progress Award. To date, he has published over 110 academic papers, accumulating more than 4,100 citations. He has applied for over 30 invention patents, with 19 already granted. Additionally, he has contributed to the compilation of three national standards and two industry standards.

Dong Guo

Dong Guo is a Post-Doctoral Research Fellow at the School of Civil Engineering and Transportation, Guangzhou University. He obtained his doctoral degree from The Hong Kong Polytechnic University. His research is dedicated to establishing accurate and standardized methodologies for investigating the bond behavior at the interface between fiber-reinforced polymers (FRPs) and steel or concrete substrates. A central focus of his work is understanding the resilience and long-term performance of these critical interfaces under harsh environmental conditions, such as extreme temperatures and humidity. Dr. Guo employs a multiscale approach, spanning from the molecular level of adhesive materials to the macro-scale of structural components, to elucidate underlying deterioration mechanisms. Based on these insights, he develops targeted modification techniques for the bonding interface to enhance its mechanical strength and long-term durability. His work aims to bridge the gap between fundamental material science and the reliable implementation of FRP strengthening systems in civil infrastructure.

Jun He

Jun He is an Associate Professor within the School of Civil Engineering, Chang'an University. He obtained his doctoral degree from The Harbin Institute of Technology. His primary work is dedicated to advancing the application of high-performance materials such as fiber-reinforced polymers (FRPs) in infrastructure, aiming to make engineering structures safer, more durable, and lighter. To this end, he has systematically conducted fundamental research on FRP–concrete/steel interfacial behavior, FRP-strengthened beams, FRP-strengthened columns, and FRP-strengthened arches. His research outcomes have been applied to major infrastructure projects in China. Additionally, he has developed a strong interest in the utilization of solid waste, such as molybdenum tailing concrete.

Bai Zhang

Dr. Bai Zhang is an Associate Professor within the School of Civil and Environmental Engineering at Changsha University of Science and Technology. He holds a Ph.D. in Civil Engineering from Southeast University and was recognized in the 2025 "World's Top 2% Scientists" list. His research centers on FRP-reinforced geopolymer marine concrete structures, functionalized low-carbon geopolymer concrete, and data-driven methods for composites and structural systems. Dr. Zhang has led eight research projects, including grants from the National Natural Science Foundation of China and the Hunan Provincial Natural Science Foundation. He has authored more than 80 papers in leading international civil engineering journals, including 74 SCI-indexed articles and 7 ESI Highly Cited Papers. His publications have accumulated over 2,500 citations (Google Scholar), with a highest single-article citation count of 240 and an H-index of 30. His academic contributions have been honored with awards such as the "Excellent Doctoral Dissertation of Jiangsu Province" and the "First Prize of Science and Technology of China Highway Society". Additionally, he serves on the Editorial Boards of several international journals, including *Corrosion Communications*, *Scientific Reports*, *Green Carbon*, *Energy & Environment Nexus*, *Sound & Vibration*, *Sustainable Marine Structures*, and *Discover Civil Engineering*.

Preface

The global civil engineering sector stands at a critical juncture, tasked with extending the service life of aging infrastructure while constructing new assets that are resilient, durable, and sustainable. It is within this context that advanced composite materials have emerged as a pivotal enabling technology. This Reprint, "Composite Materials for Civil Engineering Applications," is motivated by the urgent need to bridge the gap between the promising properties of these materials and their reliable, codified application in real-world structures. Our primary aim is to present a curated collection of research that not only showcases material performance but also addresses the multifaceted challenges of integration, durability, and environmental impact.

This volume is specifically addressed to a diverse audience of academic researchers, practicing civil and structural engineers, advanced students, and policymakers who are engaged in the future of construction materials. We have sought to compile works that speak to both the fundamental understanding of material behavior and the practical imperatives of construction. The scope is intentionally broad, reflecting the interdisciplinary nature of the field, and encompasses the mechanical enhancement of structures with fiber-reinforced polymers, the development of low-carbon geopolymer systems, and the innovative valorization of industrial and agricultural by-products.

A central challenge that this Reprint engages with is the transition of composites from laboratory prototypes to mainstream engineering solutions. This requires a profound understanding of long-term performance under complex environmental exposures, the refinement of interfacial interactions in hybrid systems, and the establishment of robust design guidelines. The contributions herein tackle these issues head-on, offering insights from the microstructural to the structural scale. It is our sincere hope that the scientific findings and engineering methodologies presented will serve as a valuable resource and a catalyst for innovation, ultimately contributing to the development of safer and more sustainable infrastructure worldwide.

Yanshuai Wang, Dong Guo, Jun He, and Bai Zhang
Guest Editors

Journal of
Composites Science

MDPI

Editorial

Editorial: Composite Materials for Civil Engineering Applications

Dong Guo [1], Yanshuai Wang [2,*], Bai Zhang [3] and Jun He [4]

1 School of Civil Engineering and Transportation, Guangzhou University, Guangzhou 510180, China; dong.gzhu@gzhu.edu.cn
2 College of Civil and Transportation Engineering, Shenzhen University, Shenzhen 518060, China
3 School of Civil Engineering, Changsha University of Science and Technology, Changsha 410205, China
4 School of Civil Engineering, Chang'an University, Xi'an 710064, China
* Correspondence: yswang@szu.edu.cn

Composite materials are attracting increasing attention in civil engineering due to their superior properties, such as high stiffness, excellent strength-to-weight ratios, corrosion resistance, and design tailorability. Concurrently, the global imperative for more durable, resilient, and sustainable infrastructure is driving the innovative adoption of fibre-reinforced polymers (FRPs) and other advanced composites in both new construction and the strengthening of ageing structures [1–3]. The successful integration of composites into mainstream practice necessitates addressing critical challenges, including understanding the working mechanism of FRP in various structural systems and long-term performance under complex environmental exposures, aiming at conducting full lifecycle assessments and developing reliable design codes [4,5]. This Special Issue brings together the latest research to provide a comprehensive platform bridging materials science, structural engineering, and practical implementation. This Special Issue of *Journal of Composites Science*, comprising nine research articles, contributes to this endeavour by presenting scientific insights and technological advancements in innovative construction techniques and sustainable civil engineering materials. The studies collectively address critical fronts: enhancing the durability and load-bearing capacity of structures using FRPs, developing novel low-carbon geopolymer systems, and valorising industrial and agricultural by-products for resource-efficient construction. These works demonstrate how interdisciplinary research integrating mechanics, materials science, and environmental engineering can transform conventional practices towards more resilient and sustainable infrastructure. The topics span from the dynamic behaviour of confined composites and the interface analyses of strengthened members to the microstructural design of geopolymers and the functional use of waste-derived components. Together, these papers exemplify the ongoing innovation aimed at solving pressing engineering challenges while adhering to sustainability principles. Below is a thematic overview of the featured studies.

Feng et al. [6] investigated the dynamic compressive behaviour of carbon fibre-reinforced polymer (CFRP)-confined high-water material under impact loading. Utilising a Split Hopkinson Pressure Bar (SHPB) and Scanning Electron Microscopy (SEM), they systematically analysed the effects of the water-to-cement ratio, number of CFRP layers, and strain rate. The research found that CFRP confinement significantly increased peak strength and impact resistance, refined failure modes, and mitigated microstructural damage by restraining lateral deformation. This provides a complete theoretical basis from macro to micro perspectives for designing impact-resistant support systems in deep mining.

The confinement effects of FRP are mainly manifested in the contact-critical applications of FRP-strengthened structures, while for bond-critical applications, its performance relies on the bond behaviour between the FRP and another adherend. Li et al. [7] systematically studied the coupled effects of hygrothermal environments and interface roughness on the bond behaviour and flexural performance of prestressed CFRP-strengthened cracked steel beams. Four-point bending test results revealed that high-grade sandblasting (Sa3) significantly enhanced interfacial bond strength via a synergistic "mechanical interlock–adhesion" mechanism, while end anchorage and prestress application effectively suppressed CFRP debonding and increased material utilisation efficiency to 96.8–98.5%. Although hygrothermal exposure accelerated interfacial deterioration, its influence on the ultimate load was relatively limited, offering valuable scientific insights for the design and interface treatment of steel bridges in hygrothermal regions.

Beyond the external bonding strengthening of the sole FRP composites at the circumference or tensile soffit of existing structures, they can be directly embedded within concrete matrices as reinforcements or grids, forming Carbon Fabric-Reinforced Cementitious Matrix (CFRCM) strengthening overlays. Zhu et al. [8] studied the degradation of tensile properties in CFRCM composites under electrochemical anodic polarisation, comparing the effects of three standard electrolyte solutions (NaCl, NaOH, simulated concrete pore solution). The results indicated that anodic polarisation significantly reduced the peak tensile strength and post-cracking stiffness, with the degradation severity dependent on the electrolyte type. Crucially, it was found that strength degradation was more pronounced for carbon-fibre bundles embedded in the mortar matrix than for bare bundles, providing critical durability data for integrated cathodic protection–structural strengthening systems.

In the analysis of novel composite structures, Çelik al. [9] investigated the free vibration characteristics of functionally graded material (FGM) Timoshenko beams using both analytical and numerical approaches. They derived governing equations using Hamilton's principle and solved them using the Navier solution method, while also establishing a 3D finite element model in ABAQUS for verification. The results showed a high level of agreement between the two methods. This research provides a reliable methodology for the dynamic analysis of advanced composite structures with complex material gradients.

Beyond the structural usage of the composite materials, the philosophy of sustainable development drives construction materials towards low carbonisation and waste resource utilisation, a trend not only evident in superstructures but also extending to geotechnical engineering. In the development of low-carbon cementitious materials, geopolymers are a promising alternative to ordinary Portland cement (OPC). Pourdolat et al. [10] innovatively employed silica fume-derived sodium silicate (SSA) combined with NaOH as an activator for fly ash-based geopolymer mortar. The study showed that the SSA system exhibited comparable workability and longer setting times than the commercial sodium silicate (SSC) system, while achieving superior mechanical properties and a denser microstructure when cured at 65 °C and 80 °C. Economic and environmental assessments indicated an approximate 30% cost reduction and a modest decrease in carbon emissions. This work demonstrates the technical viability of SSA as a sustainable activator for next-generation geopolymers. To further improve the carbonation resistance of geopolymer material, Li et al. [11] systematically examined the effects of the magnesium oxide (MgO) content and metakaolin-to-fly ash ratio on the carbonation performance, mechanical properties, and microstructure of geopolymer pastes. The research found that MgO addition significantly enhanced compressive strength and carbonation resistance, reducing the 28-day carbonation depth by up to 80.4%. Microstructural analysis indicated that MgO promoted the hydration process, forming fine $Mg(OH)_2$ crystals that refined the pore structure and increased matrix densification.

The composite materials are also applied to improve the properties of traditional cementitious materials. Yu et al. [12] introduced basalt fibre into foam concrete and pro-

https://doi.org/10.3390/jcs10020067

posed a novel methodology based on Digital Image Correlation (DIC) technology and the DBSCAN clustering method to quantitatively analyse tensile damage evolution and localisation behaviour. The statistical results indicated that basalt fibre incorporation not only effectively delayed the damage process and increased the initial damage threshold load of foam concrete but also significantly inhibited damage localisation. This provides an advanced characterisation tool for studying the damage mechanisms and optimising the performance of such lightweight, high-strength materials.

In addition, Amorim et al. [13] investigated the feasibility of incorporating industrial rolling mill residues (lamination waste) into mortars. The chemical, physical, microstructural, and mechanical properties were evaluated using various techniques including XRF, XRD, FTIR, and SEM/EDS. The results showed that incorporating industrial lamination waste could improve the mechanical strength of mortars and reduce capillary water absorption in some mixes, proving it a technically feasible and environmentally friendly alternative.

In geotechnical engineering, composites are used to modify soil properties. Cheng et al. [14] focused on the influence of rubber fibre on the engineering characteristics of calcareous sand. A series of triaxial tests revealed that increasing the rubber fibre content reduced the strength of rubber–calcareous sand but effectively suppressed dilatancy, reduced particle breakage, and altered its permeability. The study established a power-function relationship between the input energy and the relative breakage rate of calcareous sand, providing a scientific basis for selecting the optimal mix ratio for the environmentally friendly use of waste rubber in road subgrade fill engineering.

The research compiled in this Special Issue comprehensively demonstrates the latest advances in composite materials for addressing key challenges in civil engineering, from enhancing the dynamic performance and durability of FRP-strengthened structures [6–8] and FGM beams [9] to developing novel geopolymers with high performance and durability [10,11], utilising fibres [12] or industrial waste [13,14] to modify construction materials. These works collectively emphasise the core importance of multi-scale understanding (from microstructure to macro-performance), analysis of multi-factor coupling effects, and sustainability assessment in modern composite material.

Despite significant progress, the path to the widespread application of composites in civil engineering remains fraught with challenges. Future research should focus more on establishing long-term performance databases and predictive models for materials under complex multi-field coupled environments, developing performance-based design theories and codified standards, and exploring intelligent monitoring and lifecycle management technologies for composite structures.

We believe that the outcomes presented in this Special Issue will serve as a valuable reference for researchers, engineers, and students in the field and inspire further innovative work to collectively advance the role of composite materials in building safer, more durable, and more sustainable infrastructure.

Conflicts of Interest: The authors declare no conflicts of interest.

References

1. Guo, D.; Zhu, M.-C.; Deng, J.; Zhong, M.-T.; Zhou, H. A novel methodology for determining the FRP-to-steel/concrete bond-slip relationship from load-displacement curves under thermal effects. *Compos. Struct.* **2025**, *377*, 119886. [CrossRef]
2. Wang, Y.-S.; Alrefaei, Y.; Dai, J.-G. Influence of coal fly ash on the early performance enhancement and formation mechanisms of silico-aluminophosphate geopolymer. *Cem. Concr. Res.* **2020**, *127*, 105932. [CrossRef]
3. Guo, D.; Liu, Y.-L.; Gao, W.-Y.; Dai, J.-G. Bond behavior of CFRP-to-steel bonded joints at different service temperatures: Experimental study and FE modeling. *Constr. Build. Mater.* **2023**, *362*, 129836. [CrossRef]

https://doi.org/10.3390/jcs10020067

4. Wang, Y.-S.; Peng, K.-D.; Alrefaei, Y.; Dai, J.-G. The bond between geopolymer repair mortars and OPC concrete substrate: Strength and microscopic interactions. *Cem. Concr. Compos.* **2021**, *119*, 103991. [CrossRef]
5. Dong, Z.; Han, T.; Zhang, B.; Zhu, H.; Wu, G.; Wei, Y.; Zhang, P. A review of the research and application progress of new types of concrete-filled FRP tubular members. *Constr. Build. Mater.* **2021**, *312*, 125353. [CrossRef]
6. Feng, F.; Meng, S.; Huang, H.; Zhou, Y.; Zhao, H. Dynamic Compressive Behavior of CFRP-Confined High Water Material. *J. Compos. Sci.* **2025**, *9*, 482. [CrossRef]
7. Li, J.; Wu, K.; Yang, M. Influence of Interface Roughness and Hygrothermal Environment on the Flexural Performance of Prestressed CFRP-Strengthened Cracked Steel Beams. *J. Compos. Sci.* **2025**, *9*, 602. [CrossRef]
8. Zhu, M.; Zhang, Y.; Chen, H.; Deng, J.; Zeng, C. Degradation of Tensile Properties in CFRCM Composites Under Anodic Polarization: Role of Standardized Electrolyte Solutions. *J. Compos. Sci.* **2026**, *10*, 16. [CrossRef]
9. Çelik, M.; Demirkan, E.; Yıldırım, A.F. Analytical and Numerical Investigation of Vibration Characteristics in Shear-Deformable FGM Beams. *J. Compos. Sci.* **2025**, *9*, 567. [CrossRef]
10. Pourdolat, N.; Raizada, P.; Gupta, R. Mechanical and Microstructural Performance of Fly Ash-Based Geopolymer Mortar Activated by Silica Fume-Derived Sodium Silicate. *J. Compos. Sci.* **2026**, *10*, 22. [CrossRef]
11. Li, S.; Ji, D. Tuning the Carbonation Resistance of Metakaolin–Fly Ash-Based Geopolymers: The Dual Role of Reactive MgO in Microstructure and Degradation Mechanisms. *J. Compos. Sci.* **2025**, *9*, 549. [CrossRef]
12. Yu, H.; Liu, C.; An, Y.; Ma, R.; Liu, Y. Statistical Analysis of Tensile Damage of Basalt Fiber Foam Concrete Based on DBSCAN Clustering Method. *J. Compos. Sci.* **2025**, *9*, 694. [CrossRef]
13. Amorim, A.L.M.; Oliveira, J.V.B.L.; Coelho, R.C.M.; Teti, B.S.; Costa, E.C.; Lima, N.B.; Alves, K.G.B.; Lima, N.B.D. Sustainable Mortars Incorporating Industrial Rolling Mill Residues: Microstructural, Physical, and Chemical Characteristics. *J. Compos. Sci.* **2026**, *10*, 42. [CrossRef]
14. Cheng, Y.; Geng, H.; Wang, L.; Wang, Y.; Yang, G.; Xie, Y.; Ma, L.; Li, C. Effect of Rubber Fiber Content on the Mechanical Properties of Calcareous Sand. *J. Compos. Sci.* **2025**, *9*, 578. [CrossRef]

Disclaimer/Publisher's Note: The statements, opinions and data contained in all publications are solely those of the individual author(s) and contributor(s) and not of MDPI and/or the editor(s). MDPI and/or the editor(s) disclaim responsibility for any injury to people or property resulting from any ideas, methods, instructions or products referred to in the content.

https://doi.org/10.3390/jcs10020067

Journal of
Composites Science

Article

Dynamic Compressive Behavior of CFRP-Confined High Water Material

Feiyang Feng [1], Shuling Meng [2], Haishan Huang [3], Yafei Zhou [4] and Hongchao Zhao [5,*]

1 School of Geology and Mining Engineering, Xinjiang University, Urumqi 830047, China; 107552404778@stu.xju.edu.cn
2 China West Construction Group Co., Ltd., Urumqi 610200, China; 15037445129@163.com
3 Xinjiang Saier Energy & Mining Co., Ltd., Tacheng 834406, China; 18999317182@163.com
4 Korla Jinchuan Mining Co., Ltd., Korla 841001, China; zhouyafei25400@163.com
5 State Key Laboratory of Intelligent Construction and Healthy Operation and Maintenance of Deep Underground Engineering, Xuzhou 221116, China
* Correspondence: zhaohongchao@xju.edu.cn

Abstract: As mining operations extend deeper underground, support structures are increasingly subjected to severe impact loads. The dynamic mechanical performance of column-type support systems has, therefore, become a pressing concern. In the present research, a Split Hopkinson Pressure Bar (SHPB) apparatus, combined with Scanning Electron Microscopy (SEM), is used to systematically examine how the water-to-cement ratio, number of carbon-fiber reinforced polymer (CFRP) layers, and strain rate influence the dynamic compressive behavior and microstructural evolution of CFRP-confined high-water material. The results indicate that unconfined specimens are strongly strain rate-dependent, with peak strength following a rise–fall trend. A lower water–cement ratio results in a denser internal structure and improved strength. Additionally, CFRP confinement markedly enhances peak strength and impact resistance, refines failure modes, and promotes the formation of denser hydration products by limiting lateral deformation. This confinement effect effectively mitigates microstructural damage under high strain rates. These findings clarify the reinforcement mechanism of CFRP from both macroscopic and microscopic perspectives, offering theoretical insights and engineering references for the design of impact-resistant support systems in deep mining applications.

Keywords: underground mines; composite structures; SHPB; CFRP; high-water material

1. Introduction

As shallow coal resources become increasingly depleted, mining activities are rapidly shifting to greater depths. This transition has led to an increase in rockburst occurrences, which now pose a significant threat to the safe operation of coal mines [1–3]. Current prevention strategies primarily involve multi-parameter precursor identification systems based on acoustic emission monitoring and other sensing technologies [4–7]. These are often combined with pressure relief techniques such as large-diameter borehole drilling [8–12], alongside roadway support methods like bolt–mesh systems for surrounding rock control [13–15]. Together, these approaches form an integrated "early warning–pressure relief–support" system for rockburst mitigation.

To further reduce the damage caused by rockbursts, researchers have proposed various novel energy-absorbing support systems based on different reinforcement strategies.

 https://doi.org/10.3390/jcs9090482

Xu et al. [16] innovatively introduced a steel honeycomb sandwich panel as an energy-absorbing device to mitigate the impact energy from rockbursts. Zhang et al. [17] investigated the influence of joint roughness coefficient and interface morphology at the rock–shotcrete interface on the dynamic splitting behavior of rock–concrete composites. Fan et al. [18] developed a novel expansion–friction composite energy-absorbing anchor cable to address the vulnerability of conventional anchors under rockburst conditions in coal mines, significantly improving their impact resistance.

In addition to these reinforcement strategies, vertical support systems are also essential components of underground mining support technology [19,20]. Batchler [21] demonstrated that the mechanical performance of pumpable column supports is jointly influenced by external confinement and the properties of internal filling materials. In underground mine support, high-water quick-setting materials have attracted considerable attention due to their strong water affinity, high fluidity, rapid setting, and ease of application. Meanwhile, fiber-reinforced composite materials have been widely adopted by researchers owing to their high specific strength and excellent corrosion resistance. To address the limitations of conventional FRP-confined columns in accommodating large deformations in mining environments, Yu et al. [22] proposed a novel composite column structure composed of an outer FRP tube and an inner filling of coal gangue–calcium sulfoaluminate-based high-water cementitious material. Building upon this design, Zhao et al. [23] further investigated the effects of fiber type, number of confinement layers, and water–cement ratio on the static mechanical properties of FRP-confined high-water material. Their findings showed that FRP provides lateral confinement during compression, with the degree of confinement determined by the type and thickness of the FRP. Stronger confinement enhances both compressive strength and deformability. The drainage behavior of high-water material also contributes to its large-deformation capacity. Liu et al. [24] conducted triaxial compression tests to simulate lateral confinement and systematically studied the compressive behavior and bleeding mechanism of high-water material under varying confining pressures. The results revealed that the seepage threshold is most significantly influenced by the water–cement ratio, followed by confining pressure and curing time. The material initially undergoes slow volumetric shrinkage, followed by rapid compression and progressive densification. A lower water–cement ratio results in reduced free water content, requiring higher confinement to prevent damage caused by water migration. However, these studies have primarily focused on static conditions, and the dynamic mechanical behavior of such materials under high strain rates remains largely unexplored.

Numerous studies have investigated the dynamic mechanical behavior of fiber-reinforced confined concrete. Yang et al. [25] conducted Split Hopkinson Pressure Bar (SHPB) tests to examine the effects of different numbers of AFRP wrapping layers and strain rates on the dynamic response of concrete. Their results indicated a positive correlation between dynamic strength and strain rate, with AFRP-confined concrete exhibiting significantly higher strength and toughness than unconfined concrete. Xiong et al. [26] evaluated the compressive behavior of CFRP-confined concrete under high-strain rates and compared it with unconfined specimens. While unconfined concrete was highly sensitive to strain rate, showing increased strength with increasing strain rate, the confined concrete exhibited improved overall strength but less strain rate sensitivity. Guo et al. [27] developed a unified model for the dynamic compressive strength of FRP-confined concrete based on an experimental database. Under certain FRP confinement conditions, both the compressive strength and corresponding strain of unconfined and FRP-confined concrete increased with strain rate. Moreover, as the FRP confinement ratio increased, the dynamic enhancement factor also rose. However, the confinement effectiveness of FRP is more prominent under quasi-static loading. Under dynamic loading, concrete may fail before the FRP confinement is fully mobilized, resulting in a non-linear relationship between confinement ratio and enhancement efficiency. Jiang et al. [28] identified a typical three-stage stress–strain behavior of FRP-confined concrete under dynamic compression: initial rise, post-peak decline, and secondary rise. Their experiments confirmed this pattern and revealed that the lateral–axial strain curve under dynamic loading is similar in shape to that under quasi-static conditions, though the confinement effect of BFRP is more pronounced in the dynamic regime. Strain rate history analysis showed asynchronous changes between axial and lateral strain rates near the peak, leading to a lower dynamic confinement ratio compared to the static case and contributing to the post-peak decline. To characterize this effect, the concept of dynamic confinement ratio (fl,d/fcd) was introduced, with a threshold value of 0.045 proposed to ensure effective confinement. While these studies provide a solid theoretical foundation for the present work, the dynamic behavior of high-water backfill materials confined by fiber-reinforced composites remains largely unexplored.

Building upon the above, this study focuses on the dynamic behavior of CFRP-confined high-water material. A series of Split Hopkinson Pressure Bar (SHPB) impact compression tests was conducted, accompanied by Scanning Electron Microscopy (SEM) analysis. Both unconfined and CFRP-confined groups were tested under varying conditions, including different water–cement ratios (1.25, 1.5, 1.75), numbers of CFRP confinement layers (0, 1, 3), and strain rates. The experiments systematically examined the effects of water–cement ratio, confinement condition, and strain rate on the dynamic compressive strength and microstructural characteristics of CFRP-confined high-water material.

2. Experimental Program

2.1. Test Specimens

The specimens were primarily composed of high-water material, water, adhesive, and carbon fiber-reinforced polymer (CFRP) sheets. The high-water material, supplied by angzhou Zhongkuang Building New Materials Technology Co., Ltd., Yangzhou, China, consisted of two separate components: Component A, composed mainly of calcium aluminate and calcium sulfoaluminate, and Component B, primarily composed of gypsum and lime. Tap water was used for mixing. The adhesive and CFRP sheets, responsible for bonding and confinement, respectively, were provided by Shanghai Hanma Construction Technology Co., Ltd., Shanghai, China. The tensile properties of the CFRP are listed in Table 1 [29]. The adhesive was prepared by mixing carbon fiber impregnating resin (HM-180C3P, Part A) and structural epoxy resin (HM-120CP, Part B) at a ratio of 2:1.

Table 1. Mechanical Properties of CFRP.

Material	Average Tensile Strength (MPa)	Average Strain ($\times 10^{-2}$)	Elastic Modulus (GPa)
CFRP	3346.89	1.23	267.57

As shown in Table 2, a total of 41 cylindrical specimens were prepared for the experiment, including 20 unconfined specimens and 21 CFRP-confined specimens. The tests were conducted using a Split Hopkinson Pressure Bar (SHPB) apparatus under impact compression loading. Grouping was performed based on the single-variable method. The water-to-binder ratio (WB) was categorized into 1.25, 1.50, and 1.75, labeled as A, B, and C, respectively. The number of CFRP layers was denoted as C0, C1, and C3. Different bullet velocity intervals were coded as "v + the lower bound of the interval." For example, the interval of 4–5 m/s was labeled as v4. Each specimen was identified by a combination of its water-to-binder ratio, CFRP layer count, and bullet velocity range. For example, specimen

A-C0-v2 refers to a sample with a water-to-binder ratio of 1.25, no CFRP confinement, and a bullet velocity range of 5–6 m/s.

Table 2. Grouping and Number of Specimens.

<table>
<tr><th>Group Label</th><th>Bullet Velocity (m/s)</th><th>Water–Cement Ratio</th><th>Number of CFRP Layers</th><th>Number of Specimens</th></tr>
<tr><td>B-C0-v4</td><td>4~5</td><td rowspan="7">1.50</td><td rowspan="3">0</td><td>4</td></tr>
<tr><td>B-C0-v5</td><td>5~6</td><td>4</td></tr>
<tr><td>B-C0-v6</td><td>6~7</td><td>4</td></tr>
<tr><td>B-C3-v9</td><td>9~10</td><td rowspan="4">3</td><td>3</td></tr>
<tr><td>B-C3-v10</td><td>10~11</td><td>3</td></tr>
<tr><td>B-C3-v11</td><td>11~12</td><td>3</td></tr>
<tr><td>B-C3-v12</td><td>12~13</td><td>3</td></tr>
<tr><td>A-C0-v5</td><td rowspan="2">5~6</td><td>1.25</td><td rowspan="2">0</td><td>4</td></tr>
<tr><td>C-C0-v5</td><td>1.75</td><td>4</td></tr>
<tr><td>A-C3-v12</td><td rowspan="2">12~13</td><td>1.25</td><td rowspan="2">3</td><td>3</td></tr>
<tr><td>C-C3-v12</td><td>1.75</td><td>3</td></tr>
<tr><td>B-C1-v9</td><td>9~10</td><td>1.50</td><td>1</td><td>3</td></tr>
</table>

The specimen preparation procedure was as follows. Components A and B, along with two portions of water, were weighed according to the designed mix proportions. Component A was first mixed thoroughly with one portion of water, and Component B was mixed with the other portion. The two slurries were then rapidly combined and stirred immediately, given the rapid setting behavior after their mixing. Once mixed, the slurry was quickly poured into cylindrical molds with a diameter of 50 mm and a height of 100 mm to form the matrix.

After complete setting, the specimens were demolded and wrapped in plastic film for natural curing. After 3 days of curing, CFRP sheets were prepared for confinement. Sheets with a width of 10 cm and lengths of 20 cm and 52 cm were cut to provide one-layer and three-layer confinement, respectively. The CFRP sheets were bonded to the specimen surfaces using pre-mixed adhesive. To prevent peeling at the ends, the terminal sections were wrapped and secured with plastic film. After the adhesive was fully cured, initial confinement was complete.

To meet the height-to-diameter ratio requirements of the Split Hopkinson Pressure Bar (SHPB) test—recommended to be between 0.5 and 1.0 according to the GB/T 34108-2017 standard [30]—all specimens were machined to a standard size of 25 mm in height and 50 mm in diameter. This ensured consistency in testing conditions and comparability of the results. The detailed specimen preparation process is illustrated in Figure 1.

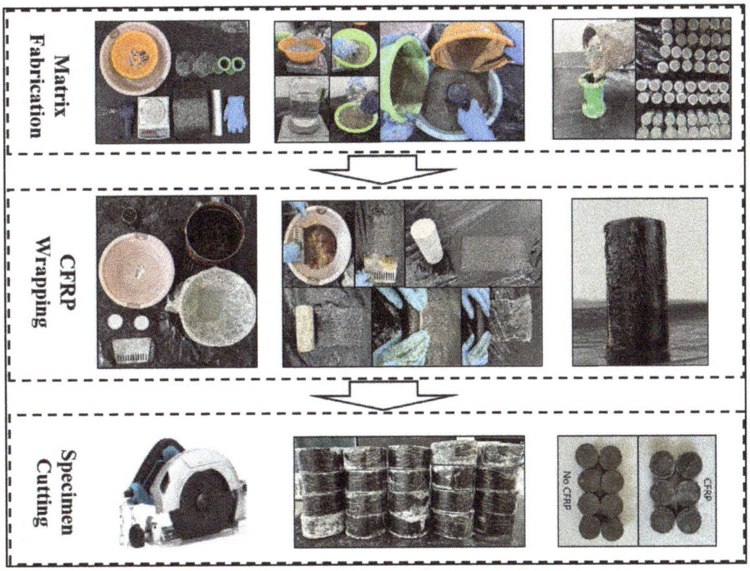

Figure 1. Specimen Preparation Flowchart.

2.2. Experimental Equipment and Procedure

The dynamic loading tests were conducted using a Split Hopkinson Pressure Bar (SHPB) apparatus with a bar diameter of 50 mm. A schematic diagram of the SHPB system is shown in Figure 2.

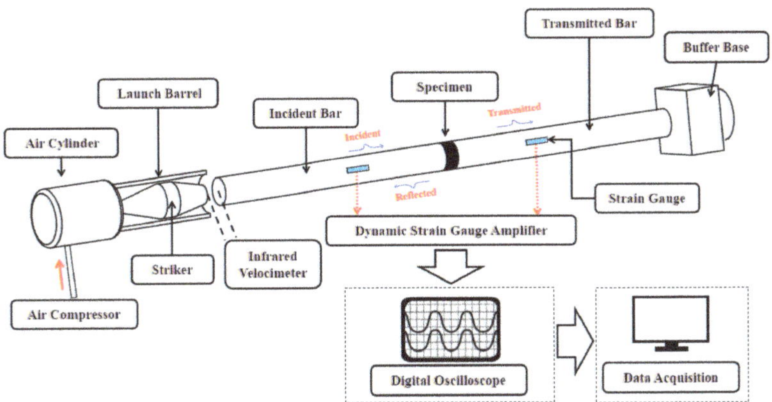

Figure 2. Schematic of the SHPB test setup.

In the SHPB tests, different strain rate conditions were achieved by adjusting the air pressure in the cylinder, which controlled the striker velocity. This, in turn, modified the shape of the one-dimensional stress wave propagating through the incident bar. When the stress wave reached the specimen, part of the wave energy was reflected into the incident bar, while the remaining portion was transmitted into the transmission bar. Strain gauges positioned between the incident and transmission bars were used to capture waveform signals, allowing the incident, reflected, and transmitted waves to be recorded separately. Based on stress wave theory and the assumption of stress equilibrium, the time-dependent strain, strain rate, and stress of the specimen were calculated using Equations (1)–(3). To ensure experimental accuracy, specimens were carefully aligned axially during testing. To minimize the influence of end-face friction on the specimen's stress state, a thin layer of petroleum jelly was applied to both ends of each specimen to reduce the friction coefficient at the specimen–bar interfaces.

$$\dot{\varepsilon}_s(t) = \left(\frac{2C_0}{L_s}\right)\varepsilon_r(t) \tag{1}$$

$$\varepsilon_s(t) = \pm\left(\frac{2C_0}{L_s}\right)^t \int_0^t \varepsilon_r(t)dt \tag{2}$$

$$\sigma(t) = \pm E\frac{A_B}{A_S}\varepsilon_t(t) \tag{3}$$

In these equations, $\dot{\varepsilon}_s(t)$ denotes the strain rate of the specimen at time t (s^{-1}), $\varepsilon_s(t)$ is the strain of the specimen, and $\sigma(t)$ is the stress (Pa). $\varepsilon_r(t)$ and $\varepsilon_t(t)$ represent the strain signals of the reflected and transmitted waves, respectively. C_0 is the wave velocity in the pressure bar (m/s), L_s is the specimen length (m), and E is the elastic modulus of the pressure bar material (Pa). A_B and A_S are the cross-sectional areas of the pressure bar and the specimen, respectively (m^2).

Figure 3 shows the incident, reflected, and transmitted waves recorded during the tests. The dynamic compression experiments were conducted at strain rates ranging from 100 to 300 s^{-1}. High-water materials often exhibit relatively high strain rates in SHPB tests, primarily due to their low strength, low elastic modulus, and high deformability. Compared to conventional concrete, high-water materials contain more free water and have a looser internal structure, making them more prone to rapid deformation under impact loading. This results in greater strain accumulation within a short period.

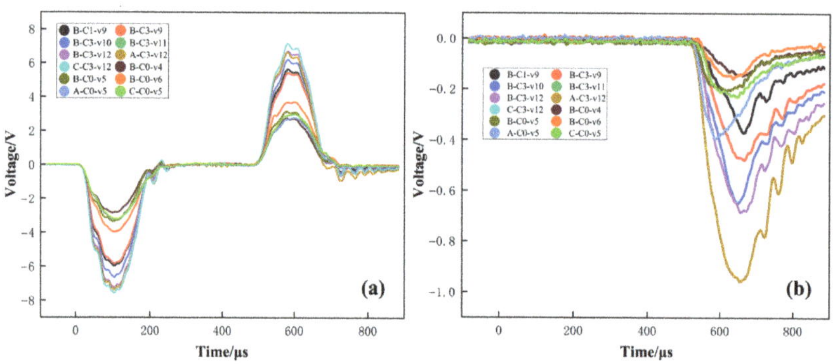

Figure 3. Typical incident, reflected, and transmitted waves in the SHPB test: (**a**) incident and reflected waves; (**b**) transmitted wave.

Moreover, the acoustic impedance of high-water material is significantly lower than that of the steel pressure bars, leading to strong wave reflections at the specimen–bar interfaces. This reduces the effective loading duration and accelerates strain accumulation. The material tends to fracture rapidly under impact, often before stress redistribution is completed, which further increases the strain rate. Therefore, even under the same loading velocity, the inherent characteristics of high-water material—such as their high water content and soft structure—cause them to respond more quickly, reaching strain rate levels significantly higher than those typically observed in concrete.

2.3. Microstructural Observation via SEM

To investigate the internal microstructure of high-water material after impact loading, scanning electron microscopy (SEM) observations were performed on the residual fragments of specimens subjected to SHPB dynamic compression tests. Representative fragments were selected from the fractured specimens, and each was split or cut along the fracture surface to expose a fresh internal cross-section suitable for observation.

The selected surfaces were then coated with a gold film approximately 20 nm thick using a vacuum sputter coater to ensure adequate electrical conductivity. Sputtering was conducted for at least 20 min to form a uniform conductive layer. After coating, the samples were placed in a high-vacuum drying chamber for more than 30 min of evacuation. Following this standard SEM preparation procedure, the fragments were mounted on specimen holders for microstructural observation.

3. Results and Discussion

3.1. Dynamic Compressive Behavior of Unconfined High-Water Material

3.1.1. Effect of Strain Rate on Peak Stress

As shown in Figure 4, at a fixed water–cement ratio of 1.50, the peak stress generally follows an initial increase followed by a decrease as the bullet velocity increases. When the velocity rises from v1tov2, the peak stress increases significantly, reaching a maximum of 9.75 MPa in the v2 range, with absorbed energy of about 9.01 J. However, further increasing the velocity to v3 results in a decline in peak stress. This reduction is attributed to rapid failure before the internal stress redistribution is completed. Under such high loading rates, microcracks initiate and propagate almost instantaneously, leading to premature macro-scale failure and a consequent drop in strength. Additionally, strain rate is positively correlated with bullet velocity: higher velocities result in higher strain rates.

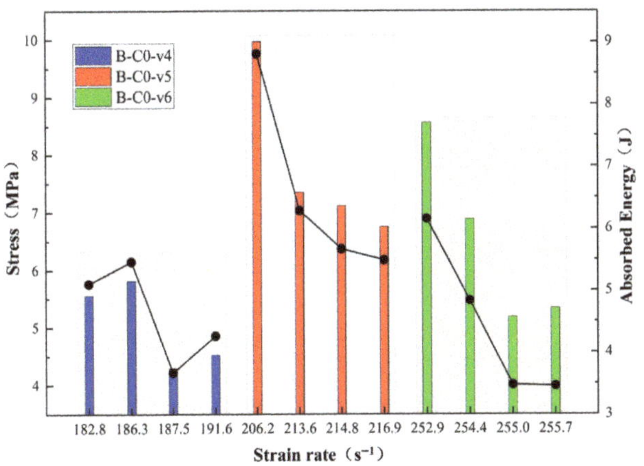

Figure 4. Strain rate–stress–energy absorption curves at different impact velocities (The line graph indicates peak stress, while the bar chart represents absorbed energy.).

3.1.2. Effect of Water–Cement Ratio on Peak Stress

As illustrated in Figure 5, under a constant bullet velocity, the peak stress decreases significantly with increasing water–cement ratio. When the ratio increases from 1.25 to 1.75, the peak stress drops from 16.75 MPa to 5.70 MPa, and the absorbed energy decreases from 14.05 J to 5.30 J. These results demonstrate that the strength of high-water material is highly sensitive to the water–cement ratio. At lower ratios, hydration products more fully fill the internal pores, resulting in a denser microstructure, higher strength, and inhibited crack propagation. In contrast, higher ratios increase the free water content and porosity after setting, making the material more porous and reducing its strength. Under unconfined conditions, the dynamic compressive strength of high-water material is highly strain-rate sensitive. It tends to increase with rising strain rate within a certain range, but when the loading rate becomes excessively high, the strength declines instead. This is because at very high strain rates, internal pores cannot close in time to bear increased stress, and rapid pulverization under impact leads to a sudden loss of strength.

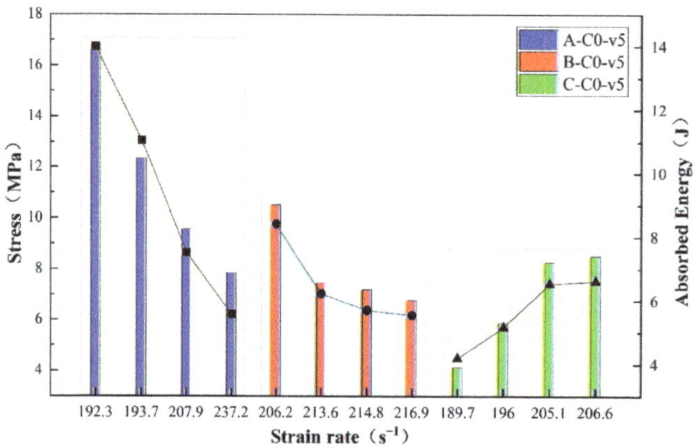

Figure 5. Strain rate–stress–energy absorption curves under different water–cement ratios (The line graph indicates peak stress, while the bar chart represents absorbed energy.).

3.1.3. Analysis of Fractal Dimension Results

To quantitatively characterize the particle size distribution of fragments after impact failure, the cumulative mass method in fractal theory was used to calculate the fractal dimension. The calculated values were statistically analyzed and are presented in Figure 5. The fractal dimension was determined using the following equation:

$$D = 3 - b \tag{4}$$

$$b = \frac{log(\frac{M_R}{M})}{logR} \tag{5}$$

In this equation, D is the fractal dimension; b is the slope of the linear fit of $log\left(\frac{M_R}{M}\right)$ versus $logR$; M_R is the cumulative mass (g) of fragments passing through a sieve with an aperture size R; and M is the total mass of the fragments (g).

Figure 6 presents the fractal dimensions of fragmented specimens under different strain rates and water–cement ratios, along with representative images of the fracture morphologies. As shown in the figure, under the same water–cement ratio, an increase in strain rate leads to more severe fragmentation. Large blocks are significantly reduced, while the proportion of fine debris increases. The corresponding fractal dimension increases from approximately 3.0 to over 3.6, indicating a shift in fracture mode from block-like failure to pulverized failure. This demonstrates that higher strain rates result in more intense specimen fragmentation, with smaller fragment sizes and a clear upward trend in fractal characteristics. Under the same impact velocity, specimens with higher water–cement ratios show a higher proportion of fine particles after failure, resulting in higher fractal dimensions. In contrast, specimens with lower water–cement ratios tend to retain larger fragments, corresponding to lower fractal dimensions. The comparison of fracture morphologies further supports this trend: specimens with a water–cement ratio of

1.25 retained relatively large and intact fragments after failure, whereas those with a ratio of 1.75 were almost completely reduced to fine particles.

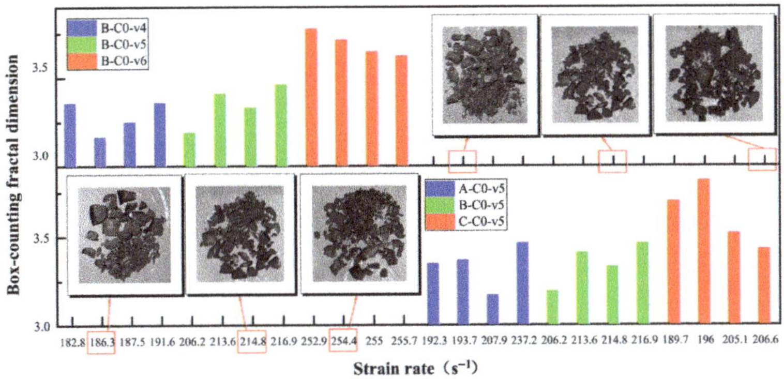

Figure 6. Fractal dimension of high-water material under unconfined conditions.

3.2. Dynamic Compressive Behavior of CFRP-Confined High-Water Material

3.2.1. Stress–Strain Curves Under Different Confinement Layers

Figure 7 compares the stress–strain curves of high-water specimens with one-layer and three-layer CFRP confinement under a bullet velocity of 9–10 m/s at a water–cement ratio of 1.50. For reference, the corresponding quasi-static stress–strain curves and typical post-impact failure cross-sections are also included. The results demonstrate that CFRP confinement significantly enhances the impact resistance of high-water material. As the number of CFRP layers increases from one to three, the peak stress rises from approximately 11 MPa to nearly 19 MPa—an increase of over 72%. The peak strain also increases, indicating that multilayer confinement improves the material's deformation capacity.

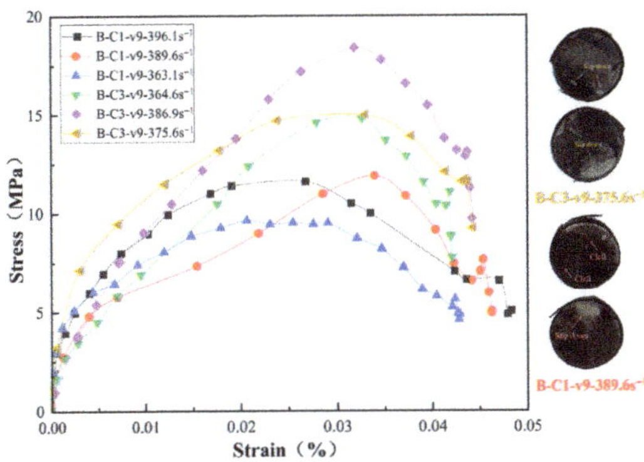

Figure 7. Stress–strain curves of CFRP-confined high-water specimens with different confinement layers under the same water–cement ratio and impact velocity.

This enhancement is consistent with previous findings on FRP-confined concrete, where external confinement induces a triaxial compression state in the core material, thereby improving both compressive strength and ductility. The observed failure morphologies further support this conclusion. The specimen with one-layer CFRP confinement exhibited edge spalling and localized fiber rupture, with scattered debris after impact. In contrast, the specimen with three-layer confinement maintained an almost intact appearance, with only minor cracks near the ends and no major structural disintegration, as shown in Figure 7. In addition, partial debonding was observed at the CFRP–high-water material interface.

This localized debonding indicates that under high strain rates, interfacial stresses can exceed the bonding strength, resulting in interfacial slip or separation in specific regions. Nevertheless, the CFRP layer as a whole remained intact and continued to provide effective lateral confinement, consistent with the overall behavior of CFRP-confined high-water materials, where confinement primarily resists lateral expansion but may still experience localized weakening under extreme loading conditions.

The dynamic stress–strain response of CFRP-confined high-water material exhibits a distinct three-stage pattern, which differs from that observed in unconfined specimens. As shown in Figure 7, the curve initially features a nearly linear rise, corresponding to the elastic deformation of the material—this is the initial rising stage. It is followed by a plateau or slight decline, indicating the onset of matrix yielding and the progressive development of internal microcracks—this is referred to as the descending stage. As the strain approaches the peak strength of the matrix, some curves exhibit a secondary increase in stress, forming a secondary rising stage.

This phenomenon arises because, once the high-water matrix begins to crack and expand under impact loading, the outer CFRP layer becomes progressively activated and provides lateral confinement, which enhances the load-bearing capacity. This delayed confinement effect leads to a secondary stress rise beyond the matrix peak, continuing until the CFRP jacket either ruptures or the impact loading concludes.

A similar three-stage dynamic response was also reported by Yang et al. [24] in SHPB tests on AFRP-confined concrete. They attributed the secondary rise to the confinement effect of the fiber jacket that becomes active after yielding of the core material.

In the present study, since the specimens were not fully destroyed by a single impact, the stress began to recover after the peak and then gradually declined—this reflects the unloading process after loading termination, as seen at the end of the curves in Figure 6.

It is worth noting that the magnitude of the secondary rising stage depends on the matrix strength and the number of CFRP layers. Stronger matrices or thicker confinement layers tend to delay severe damage at the peak, and the CFRP confinement may not be fully activated, resulting in a less prominent secondary rise. In contrast, for weaker matrices or lower confinement levels, noticeable matrix damage occurs near the peak, which triggers a more significant confinement response from the CFRP, leading to a more pronounced secondary rise. For instance, in this study, the specimen with one layer of CFRP showed a clearer stress rebound after the peak, while the specimen with three layers exhibited a less noticeable secondary rise.

3.2.2. Stress–Strain Curves at Different Impact Velocities

As shown in Figure 8, under a water–cement ratio of 1.50 and three-layer CFRP confinement, the stress–strain curves at different bullet velocities exhibit a trend similar to that of the unconfined specimens: the peak stress initially increases and then decreases with increasing impact velocity. At lower velocities, the specimens show relatively low peak stresses, indicating that the impact resistance potential of the material has not been fully activated. As the impact velocity increases to a moderate level, both the peak stress and strain capacity increase significantly, suggesting that the material strength is better mobilized under the combined effects of inertia and lateral confinement.

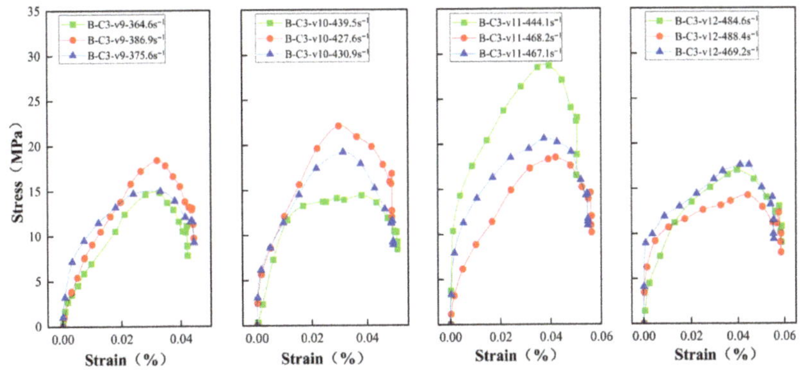

Figure 8. Stress–strain curves of CFRP-confined high-water specimens under constant water–cement ratio and confinement layers with varying impact velocities.

However, when the bullet velocity increases further, the peak stress shows a slight decline, despite continued increases in strain rate and total strain. This may be attributed to the rapid compaction of internal pores within the high-water material during instantaneous loading, which weakens the load-bearing capacity under a single high-speed impact.

3.2.3. Stress–Strain Curves Under Different Water–Cement Ratios

As shown in Figure 9, under constant impact velocity and confinement conditions, increasing the water–cement ratio from 1.25 to 1.75 results in a decrease in peak stress from over 30 MPa to approximately 24 MPa, while the peak strain increases from around 2.5% to nearly 4%. A higher water–cement ratio leads to weaker dynamic load-bearing capacity and more pronounced deformation. In contrast, specimens with lower water–cement ratios have higher solid content and more complete hydration, forming a denser internal structure. These specimens maintain higher stress levels throughout the loading process, with significantly higher peak strength than those with higher ratios.

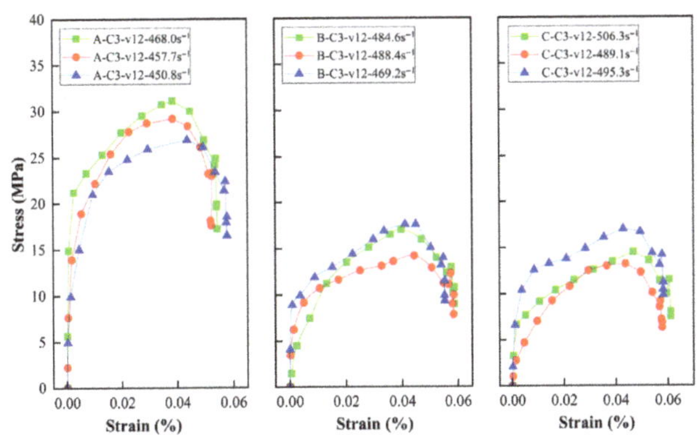

Figure 9. Stress–strain curves of CFRP-confined high-water specimens with different water–cement ratios under constant confinement layers and impact velocity.

CFRP confinement mitigates the adverse effect of increasing water–cement ratio on peak strength. Compared to unconfined specimens, the reduction in strength caused by increasing the water–cement ratio from 1.25 to 1.75 is less significant under confinement. This indicates that external confinement can partially offset the strength loss caused by increased porosity in high water–cement ratio materials, thereby enhancing the effective utilization of the internal structure.

3.2.4. Influence of Inertial Effect Under High Strain Rates

Under high-strain-rate conditions, the inertial effect plays a significant role in the dynamic behavior of high-water material. It was observed in the experiments that at high strain rates, the inertial effect becomes more pronounced, limiting the material's ability to redistribute stress, which leads to localized stress concentrations. These concentrations promote rapid crack initiation and propagation, particularly in the unconfined high-water material specimens, causing the failure mode to shift from ductile to brittle, with cracks propagating almost instantaneously. The CFRP-confined specimens, on the other hand, show delayed activation of lateral confinement, providing additional resistance during crack propagation, thereby altering the failure mode. The inertial effect also exacerbates stress concentration, especially under impact loading, where the stress in localized areas is significantly higher than in other regions, making microcracks more likely to form in the stress-concentrated areas and merge into larger cracks, ultimately accelerating material failure. Meanwhile, as the strain rate increases, the amplitude of the reflected wave also increases, leading to higher localized stress within the material, further accelerating crack formation and propagation. The inertial effect modulates the interaction between the incident and reflected waves, particularly in areas where stress has already concentrated, thereby accelerating the failure process.

3.2.5. Energy Absorption and Damage Mechanism

The energy absorbed by the material during dynamic loading can be calculated using the stress–strain integral:

$$\text{Absorbed Energy} = \int_0^{\epsilon_{\max}} \sigma(\epsilon)d\epsilon \tag{6}$$

where $\sigma(\epsilon)$ represents the stress as a function of strain, and $\epsilon_{\max}$ is the maximum strain at failure. This integral captures the total energy absorbed by the material as it deforms under loading.

The energy absorption is closely related to the material's damage mechanism. As the material undergoes deformation, part of the energy is absorbed in the form of plastic deformation, and part is used to initiate and propagate cracks. In unconfined high-water material specimens, a high absorbed energy corresponds to significant crack propagation, shifting the failure mode from ductile to brittle. In contrast, CFRP-confined specimens exhibit a more controlled energy absorption process, with less crack propagation due to the lateral confinement provided by the CFRP. This results in a higher peak stress and a more gradual failure process. The observed energy absorption values provide insight into the extent of material damage, with higher energy absorption typically indicating more extensive damage and a higher likelihood of catastrophic failure.

3.3. Microstructural Analysis

Figure 10 displays representative SEM images of the fractured microstructures of high-water material under different strain rates, water–cement ratios, and confinement conditions. Based on the SEM observations, the internal structures of the specimens were compared in terms of hydration product morphology and pore characteristics.

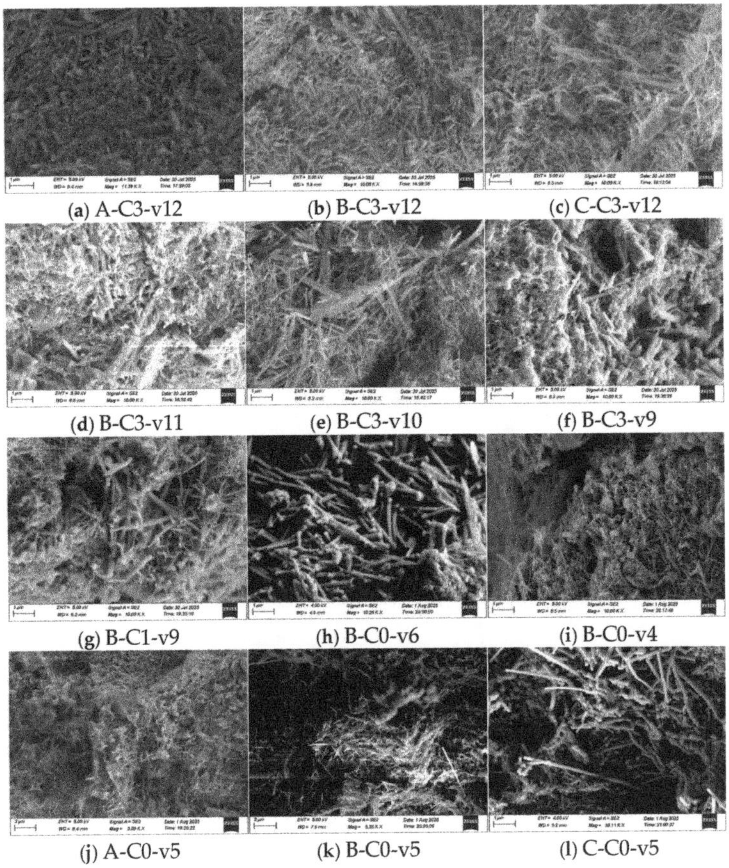

Figure 10. Microstructural Characteristics of High-Water Material.

The main hydration products of high-water material include calcium silicate hydrate (C–S–H) gel and ettringite crystals (AFt). Ettringite (i.e., calcium sulfoaluminate hydrate) exhibits distinct morphological forms under different loading conditions. According to the classification proposed by Liu et al. [24], four typical ettringite morphologies were identified in this study, as shown in Figure 11:

Type I: loose, fibrous networks;

Type II: dense fibrous structures;

Type III: short, thick rod-like crystals;

Type IV: compact granular agglomerates.

As shown in Figure 10, the unconfined specimens primarily exhibited loose fibrous networks (Type I), suggesting sparse hydration product distribution and high porosity. In contrast, CFRP-confined specimens contained a greater number of short rod-like crystals and compact granular clusters (Types III and IV), along with denser fibrous structures (Type II). This indicates that lateral confinement induces a triaxial stress state during impact, causing compaction of the hydration products. As a result, fibrous ettringite undergoes morphological transformation into denser rod-like or granular forms. The confined specimens clearly exhibited more compact microstructures with reduced porosity, whereas the unconfined specimens showed loose, highly porous structures. These observations are consistent with findings from static triaxial compression tests, in which needle-like hydration crystals tend to agglomerate and rearrange under high confining pressures, transitioning from fibrous bundles to short rods or blocky structures.

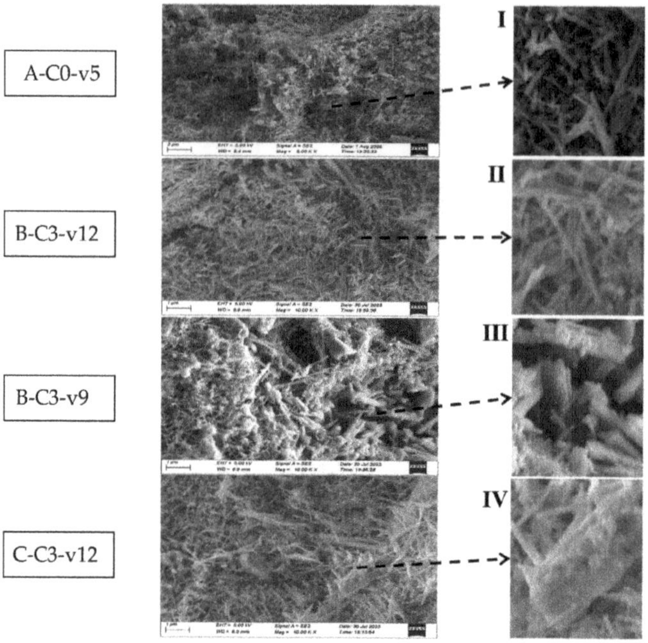

Figure 11. Four typical morphologies of ettringite observed in high-water material.

With increasing confinement levels, the morphological transformation of ettringite becomes more pronounced. In specimens with three layers of CFRP confinement, abundant rod-like crystals and dense particle clusters (Types III and IV) were observed, whereas specimens with only one layer primarily featured compacted fibrous networks (Type II). This suggests that higher confinement strength promotes compaction and transformation of hydration products during high-rate impact, significantly increasing microstructural density.

The influence of water–cement ratio on microstructure is also evident in Figure 10. Regardless of confinement condition, specimens with lower water–cement ratios exhibited more continuous and dense hydration product distributions with fewer pores. In contrast, specimens with a ratio of 1.75 showed fragmented, discontinuous microstructures and a substantial increase in visible porosity. These trends align with the macroscopic mechanical results: higher water–cement ratios imply more free water and less solid content, leading to diluted hydration and insufficient filling of voids, which results in a looser hardened structure and reduced strength. Conversely, lower ratios promote greater formation of C–S–H and ettringite, which fill the pores and create a dense matrix, thus improving strength. Therefore, reducing the water–cement ratio is an effective strategy to enhance both the microstructure and macroscopic performance of high-water material.

Under varying strain rates, the microstructural evolution of high-water material differs significantly depending on whether lateral confinement is applied. In the unconfined state, higher bullet velocities result in more extensive and larger pores and more fragmented microstructures. This is primarily due to severe radial expansion and rapid expulsion of free water under high-speed impact, which damages the original hydration framework and generates new cracks and voids. As strain rate increases, residual internal porosity increases, and the fracture surfaces become rougher and more porous. Similar findings were reported by Cao et al. [31], who used quantitative image analysis to show that, as the average strain rate increased from 46 s^{-1} to 96 s^{-1}, the average pore diameter of CTC materials increased from 6.535 μm to 31.725 μm, and the fractal dimension DDD of the pore area showed a strong positive correlation with strain rate. These results suggest that under high strain rates, the pore structure becomes more complex and widespread, corresponding to a higher degree of macroscopic fragmentation.

In contrast, CFRP-confined specimens exhibit entirely different microstructural features under impact. Due to restricted lateral expansion, a triaxial stress state forms during high-rate loading, compressing and closing internal pores. Hydration products are realigned and compacted under directed stress, preventing the formation of new voids and reducing the size of existing ones. This is clearly visible in Figure 10: unconfined specimens under high-speed impact are filled with microvoids, while their confined counterparts appear dense with no significant large pores. These observations confirm that under impact loading, lateral confinement plays a dominant role in regulating microstructural evolution. Confinement effectively suppresses internal damage propagation and helps maintain the integrity of the internal skeleton, even at high strain rates. Thus, the macroscopic strength and deformability improvements offered by CFRP confinement are largely attributed to its role in restricting lateral expansion and promoting denser hydration product accumulation.

4. Conclusions

This study investigated the dynamic mechanical behavior and microstructural evolution of high-water material confined with CFRP (Carbon Fiber Reinforced Polymer) fabrics under impact loading, using a Split Hopkinson Pressure Bar (SHPB) apparatus. The effects of water–binder ratio, confinement layers, and impact velocity (strain rate) were systematically examined. The main conclusions are as follows:

(1) Under unconfined conditions, the peak stress of high-water material shows a strain rate sensitivity: it first increases and then decreases with rising strain rate. Excessively high strain rates lead to rapid pulverization before internal stress redistribution, causing strength loss.

(2) The water–cement ratio has a significant influence on the dynamic compressive strength. A lower ratio (e.g., 1.25) results in denser hydration products, higher peak stress, and greater absorbed energy, whereas a higher ratio (e.g., 1.75) produces porous structures, reducing strength and energy dissipation capacity.

(3) The fractal dimension analysis of fracture fragments shows that both higher strain rate and higher water–cement ratio increase fragmentation degree, shifting the failure mode from block-like fracture to pulverized failure.

(4) CFRP confinement effectively enhances the impact resistance of high-water material. Increasing the number of CFRP layers raises peak stress and peak strain, restrains lateral expansion, and changes the failure mode from severe fragmentation to localized cracking.

(5) The dynamic stress–strain response of CFRP-confined high-water specimens exhibits a characteristic three-stage pattern: an initial rising stage, a descending stage, and a secondary rising stage. The secondary rise is attributed to the delayed activation of CFRP confinement after matrix cracking.

(6) At a constant confinement condition, specimens with lower water–cement ratios maintain higher peak stress and denser structures, while those with higher ratios show reduced strength but larger deformation. CFRP confinement partially offsets the strength loss caused by higher water–cement ratios.

(7) SEM analysis reveals that CFRP confinement transforms ettringite morphologies from loose fibrous forms into denser rod-like and granular structures, reducing porosity and enhancing compactness. In contrast, unconfined specimens subjected to high strain rates exhibit severe pore development and disrupted hydration frameworks.

Author Contributions: Conceptualization, F.F.; Methodology, F.F.; Funding acquisition, H.Z.; Project administration, H.Z.; Investigation, S.M.; Writing—original draft preparation, F.F.; Writing—review and editing, S.M., H.H., Y.Z. and H.Z.; Visualization, F.F. and S.M. All authors have read and agreed to the published version of the manuscript.

Funding: This study was financially supported by the National Natural Science Foundation of China (No. 52164011), Key Research and Development of Xinjiang Uygur Autonomous Region, China (2023B01010); Tianshan Talent Program of Xinjiang Uygur Autonomous Region (2023TSYCCX0095) and the State Key Laboratory of Intelligent Construction and Healthy Operation and Maintenance of Deep Underground Engineering (No. SKLGDUEK2215).

Data Availability Statement: All data generated or analyzed during this study are included in this published article. For additional information or to request the raw data, please contact the corresponding author: Hongchao Zhao, Email: zhaohongchao@xju.edu.cn.

Acknowledgments: We would like to express our thanks to the technical staff for their support during the preparation of tests.

Conflicts of Interest: Authors Shuling Meng, Haishan Huang, Yafei Zhou are employed by companies China West Construction Group Co., Ltd., Xinjiang Saier Energy & Mining Co., Ltd., and Korla Jinchuan Mining Co., Ltd.. The remaining authors declare that the research was conducted in the absence of any commercial or financial relationships that could be construed as a potential conflict of interest.

References

1. Wu, S.; Zhang, J.; Song, Z.; Fan, W.; Zhang, Y.; Dong, X.; Zhang, Y.; Kan, B.; Chen, Z.; Zhang, J.; et al. Current Status and Development of Rock Burst Disaster Prevention and Control System in China. *J. Cent. South Univ.* **2023**, *30*, 3763–3789. [CrossRef]
2. Qi, Q.; Li, H.; Deng, Z.; Zhao, Z.; Zhang, N.; Bi, Z. Research on Theory, Technology and Standard System of Rock Burst in China. *Coal Min. Technol.* **2017**, *22*, 1–5+26.
3. Dou, L.; Tian, X.; Cao, A.; Gong, S.; He, H.; He, J.; Cai, W.; Li, X. Current Status and Challenges of Rock Burst Prevention in China's Coal Mines. *J. China Coal Soc.* **2022**, *47*, 152–171.
4. Zhao, S.; Wang, E.; Wang, J.; Zhang, Q.; Li, Z.; Xu, J. Rock Burst Early Warning Utilizing Acoustic Emission and Electromagnetic Radiation Adaptive Noise Reduction and Multi-Image Feature Fusion: A Case Study. *Eng. Appl. Artif. Intell.* **2025**, *157*, 111312. [CrossRef]
5. Wang, J.; Wang, E.; Yang, W.; Li, B.; Li, Z.; Liu, X. Rock Burst Monitoring and Early Warning Under Uncertainty Based on Multi-Information Fusion Approach. *Measurement* **2022**, *205*, 112188. [CrossRef]
6. Qiu, L.; Liu, Z.; Wang, E.; He, X.; Feng, J.; Li, B. Early-Warning of Rock Burst in Coal Mine by Low-Frequency Electromagnetic Radiation. *Eng. Geol.* **2020**, *279*, 105755. [CrossRef]
7. Di, Y.; Wang, E.; Huang, T. Identification Method for Microseismic, Acoustic Emission and Electromagnetic Radiation Interference Signals of Rock Burst Based on Deep Neural Networks. *Int. J. Rock Mech. Min. Sci.* **2023**, *170*, 105541. [CrossRef]
8. Xu, P.; Shao, J.; Fan, D.; Chang, J.; Zhang, N. Analysis of Pressure Relief Effect of Borehole in Rock Burst Mine. *Energy Rep.* **2022**, *8* (Suppl. S2), 156–161. [CrossRef]
9. Gu, S.; Wang, C.; Li, W.; Gui, B.; Jiang, B.; Ren, T.; Xiao, Z. Technical Management Practice of Rock Burst Prevention and Control: A Case Study of Yankuang Energy Group Co., Ltd. *Geohazard Mech.* **2024**, *2*, 225–235. [CrossRef]
10. Yang, Z.; Dou, L.; Liu, C.; Xu, M.; Lei, Z.; Yao, Y. Application of High-Pressure Water Jet Technology and the Theory of Rock Burst Control in Roadway. *Int. J. Min. Sci. Technol.* **2016**, *26*, 929–935. [CrossRef]
11. Hao, X.; Sun, X.; Tang, Z.; Hu, P.; Luo, J.; Shen, L.; Tang, K. Source Prevention and Control Technology System and Application of "Artificial Pre-Split Layer" with High-Position Whole-Layer Blasting of Overburden. *J. China Coal Soc.* **2024**, *49*, 1318–1331.
12. Wang, Y.; Zhang, C.; Xu, Z. Study on Surrounding Rock Control Technology of Along-Goaf Roadway with Deep-Hole Blasting Roof-Cutting in Thick Coal Seam. *Min. Res. Dev.* **2023**, *43*, 103–111.

13. Chen, B.; Zuo, Y.; Zheng, L.; Zheng, L.; Lin, J.; Pan, C.; Sun, W. Deformation Failure Mechanism and Concrete-Filled Steel Tubular Support Control Technology of Deep High-Stress Fractured Roadway. *Tunn. Undergr. Space Technol.* **2022**, *129*, 104684. [CrossRef]
14. Liu, D.; Shan, R.; Wang, H.; Zhao, Y.; Li, Z.; Tong, X.; Wei, Y.; He, X. Research on Application of Annular Concrete-Filled Steel Tube Support in Deep Roadway. *J. Constr. Steel Res.* **2024**, *212*, 108273. [CrossRef]
15. Shan, R.; Li, Y.; Bai, H.; Sun, P.; Xiao, S.; Wu, H.; Zhao, X.; Liu, D.; Li, Z. Study on Deformation and Failure Mechanism of Semi-Coal Rock Soft Rock Roadway and Reinforcement Support Countermeasures of Anchor Cable with C-Shaped Tube. *Eng. Fail. Anal.* **2025**, *178*, 109708. [CrossRef]
16. Xu, X.; Wu, L.; Wang, H.; Wu, Y.-Q. Experimental and Numerical Study of Drop Hammer Test on Honeycomb Sandwich Panel Resistant to Rock Burst in Coal Mine Roadway. *Structures* **2024**, *65*, 106801. [CrossRef]
17. Zhang, T.; Wang, H.; Chen, M.; Niu, L.; Zhu, W. Effect of Interfacial Characteristics on Dynamic Splitting Behavior of Quasi Rock-Concrete Composite Layer: Towards Resilient Tunnel Support Against Rock Burst. *Tunn. Undergr. Space Technol.* **2025**, *155*, 106134. [CrossRef]
18. Fan, D.; Wang, A.; Dai, L.; Pan, Y.; Zhao, S.; Yu, X. Performance Evaluation of Novel Energy-Absorbing Anchor Cables with Expansion–Friction Structures for Supporting Roadways Prone to Rock Bursts. *Tunn. Undergr. Space Technol.* **2025**, *155*, 106158. [CrossRef]
19. Campoli, A.A. Selection of Pumpable Cribs for Longwall Gate and Bleeders Entries. In Proceedings of the 34th International Conference on Ground Control in Mining, West Virginia University, Morgantown, WV, USA, 29 July 2015; pp. 80–82.
20. Zhang, P.; Milam, M.; Mishra, M.; Hudak, W.J.; Kimutis, R. Requirements and Performance of Pumpable Cribs in Longwall Tailgate Entries and Bleeders. In Proceedings of the 31st International Conference on Ground Control in Mining, West Virginia University, Morgantown, WV, USA, 31 July–2 August 2012; pp. 1–11.
21. Batchler, T. Analysis of the Design and Performance Characteristics of Pumpable Roof Supports. *Int. J. Min. Sci. Technol.* **2017**, *27*, 91–99. [CrossRef]
22. Yu, T.; Zhao, H.; Ren, T.; Remennikov, A. Novel Hybrid FRP Tubular Columns with Large Deformation Capacity: Concept and Behaviour. *Compos. Struct.* **2019**, *212*, 500–512. [CrossRef]
23. Zhao, H.; Ren, T.; Remennikov, A. Standing Support Incorporating FRP and High Water-Content Material for Underground Space. *Tunn. Undergr. Space Technol.* **2021**, *110*, 103809. [CrossRef]
24. Liu, H.; Xia, Y.; Bai, J.; Cao, Z.; Zhang, Z.; Zhao, H. Tri-Axial Compressive Behavior of High-Water Material for Deep Underground Spaces. In *Deep Underground Science and Engineering*; Wiley: Hoboken, NJ, USA, 2024.
25. Yang, H.; Song, H.; Zhang, S. Experimental Investigation of the Behavior of Aramid Fiber Reinforced Polymer Confined Concrete Subjected to High Strain-Rate Compression. *Constr. Build. Mater.* **2015**, *95*, 143–151. [CrossRef]
26. Xiong, B.; Demartino, C.; Xiao, Y. High-Strain Rate Compressive Behavior of CFRP Confined Concrete: Large Diameter SHPB Tests. *Constr. Build. Mater.* **2019**, *201*, 484–501. [CrossRef]
27. Guo, Y.-C.; Xiao, S.-H.; Zeng, J.-J.; Zheng, Y.; Li, X.; Liu, F. Fiber Reinforced Polymer-Confined Concrete under High Strain Rate Compression: Behavior and a Unified Dynamic Strength Model. *Constr. Build. Mater.* **2020**, *260*, 120460. [CrossRef]
28. Jiang, Y.; Zhang, S.; Zhang, B.; Xue, G.; Wang, W. Experimental Study and Analytical Modeling on Dynamic Compressive Behavior of BFRP-Confined Concrete under High Strain Rates. *Constr. Build. Mater.* **2023**, *398*, 132515. [CrossRef]
29. Liu, H.; Cao, W.; Cao, Z.; Sun, J.; Yu, B.; Zhao, H. Influence of Stress Path on the Mechanical Behavior of Laterally Confined Coal: Laboratory Investigation. *Constr. Build. Mater.* **2025**, *466*, 140287. [CrossRef]
30. *GB/T 34108-2017*; Metallic Materials—High Strain Rate Compression Test Method at Ambient Temperature. Standardization Administration of China: Beijing, China, 2017.
31. Cao, S.; Xue, G.; Song, W.; Teng, Q. Strain Rate Effect on Dynamic Mechanical Properties and Microstructure of Cemented Tailings Composites. *Constr. Build. Mater.* **2020**, *247*, 118537.30. [CrossRef]

Disclaimer/Publisher's Note: The statements, opinions and data contained in all publications are solely those of the individual author(s) and contributor(s) and not of MDPI and/or the editor(s). MDPI and/or the editor(s) disclaim responsibility for any injury to people or property resulting from any ideas, methods, instructions or products referred to in the content.

Journal of
Composites Science

MDPI

Article

Tuning the Carbonation Resistance of Metakaolin–Fly Ash-Based Geopolymers: The Dual Role of Reactive MgO in Microstructure and Degradation Mechanisms

Shuai Li [1,2,*] and Dongyu Ji [3]

[1] Department of Architectural Engineering, Hunan Urban Construction College, Xiangtan 411100, China
[2] College of Civil Engineering, Xiangtan University, Xiangtan 411105, China
[3] Puyang Institute of Technology, Henan University, Puyang 457000, China; xiayuji2016@163.com
* Correspondence: lishuai1988648@163.com

Abstract: Geopolymers, as a novel class of low-carbon and eco-friendly cementitious material, exhibit outstanding durability and promote the resource utilization of industrial solid wastes. However, as a promising alternative to ordinary Portland cement, its susceptibility to carbonation-induced degradation may limit its widespread application. To address this challenge, this study systematically examined the effects of magnesium oxide (MgO) content and the metakaolin-to-fly ash ratio on the carbonation performance, mechanical properties, pH value, and microstructures of metakaolin–fly ash-based (MF-based) geopolymer pastes. The findings revealed that an increase in the fly ash ratio correlated with a decline in the compressive strength of MF-based geopolymer pastes. Conversely, the incorporation of MgO significantly enhanced the compressive strength, with higher fly ash ratios leading to more substantial improvements in strength. Furthermore, the addition of MgO and fly ash effectively mitigated the penetration of carbonation and the associated decrease in the pH value of the MF-based geopolymer pastes. Specifically, compared to the control group without MgO (M8F2-0%), MF-based geopolymer pastes with 4% and 8% MgO additions exhibited reductions in carbonation depth of 69.4% and 80.4%, respectively, after 28 days of carbonation, while pH values were observed to be 1.22 and 1.15 units higher, respectively. Additionally, microscopic structural analysis revealed that the inclusion of MgO resulted in a reduction in pore size, porosity, and mean pore diameter within the geopolymer pastes. This improvement was mainly attributed to the promotion of hydration processes by MgO, leading to the formation of fine $Mg(OH)_2$ crystals within the high-alkalinity pore solution, which enhances microstructural densification. In conclusion, the incorporation of MgO significantly improves the carbonation resistance and mechanical performance of MF-based geopolymers. It is recommended that future studies explore the long-term performance under combined environmental actions and evaluate the economic and environmental benefits of MgO-modified geopolymers for large-scale applications.

Keywords: metakaolin-fly ash-based geopolymer pastes; reactive MgO content; compressive strength; carbonation behavior; pH value; carbonation mechanisms

1. Introduction

Geopolymers represent a novel class of low-carbon and green materials that can be extensively manufactured using industrial solid wastes, aligning with sustainable development principles [1–4]. However, the inherent complexity of geopolymers presents significant challenges for practical application. Durability issues, particularly carbonation, constitute a major challenge limiting the long-term service performance of geopolymer concrete structures [5–8]. In concrete, reinforcing steel is passivated by the high-alkalinity pore solution, forming a protective oxide film that inhibits corrosion [9–12]. However, carbonation reduces the pore solution pH due to carbonic acid attack. When the pH value falls below a critical threshold, the pore solution can no longer sustain the passivating film, significantly increasing corrosion risk [13–15]. In chloride-laden environments, carbonation further decomposes Friedel's salt and accelerates chloride ion migration in concrete, promoting depassivation and corrosion initiation [16]. This vulnerability may be exacerbated in geopolymer concrete due to fundamental differences in hydration products compared to ordinary Portland cement (OPC) [17]. OPC hydration yields substantial $Ca(OH)_2$, which buffers against pore solution alkalinity and Ca^{2+} depletion during carbonation. Conversely, geopolymer reaction products contain negligible $Ca(OH)_2$ and rely primarily on free alkali in the pore solution to mitigate CO_2 ingress [18–21]. Consequently, geopolymer durability differs markedly from OPC [22]. Furthermore, the diversity of aluminosilicate source materials used in geopolymers results in complex systems with significant variations in hydration products and gel structures.

While the carbonation behavior of OPC is well-established, geopolymers demonstrate a heightened sensitivity to carbonation, and their underlying carbonation mechanisms remain insufficiently understood. Unlike OPC, geopolymers form gels such as N-A-S-H, C-S-H, and/or C-A-S-H gels upon hydration, with minimal $Ca(OH)_2$ formation [23–25]. Furthermore, their pore solutions contain high concentrations of alkali metal ions (Na^+ or K^+) [9,26]. Thus, the carbonation mechanisms observed in traditional OPC systems are not directly applicable to geopolymers. Robayo-Salazar et al. [27] reported high Na^+ concentrations in geopolymer pore solutions, with Na_2CO_3 and $NaHCO_3$ being primary carbonation products. This distinct pore solution chemistry fundamentally alters the carbonation process. The absence of a $Ca(OH)_2$ buffer typically results in accelerated carbonation rates. Bakharev et al. [28] demonstrated that slag-based geopolymer pastes exhibited inferior carbonation resistance versus OPC pastes, manifesting as greater strength loss and increased carbonation depth under both $NaHCO_3$ solution and high-concentration CO_2 exposure. Mechanistically, carbonation detrimentally affected geopolymers, especially high-calcium formulations. The lack of $Ca(OH)_2$ prevented protective $CaCO_3$ formation, leaving them directly vulnerable to CO_3^{2-} attack [29]. This contrast is further highlighted by Li et al. [24], who observed opposing trends in post-carbonation compressive strength between OPC paste and alkali-activated slag.

Given that carbonation risk significantly impedes the industrial adoption of geopolymers, researchers are actively exploring various admixtures to enhance carbonation resistance [30,31]. Abdalqader et al. [25] found that adding 10% MgO to fly ash–slag geopolymers effectively reduced carbonation depth. In alkaline media, MgO reacts to form finely dispersed $Mg(OH)_2$ crystals, which refine the microstructure. It can also generate layered double hydroxide (LDH) phases that provide strong CO_2 adsorption and buffering capacity, thereby significantly improving resistance to acidic degradation [32]. To address the inherent $Ca(OH)_2$ deficiency directly, Lv et al. [18] incorporated $Ca(OH)_2$ into geopolymers, successfully inhibiting carbonation depth progression and alkalinity loss; however, excessive additions markedly reduced post-carbonation compressive strength [33]. Other mineral admixtures, including silica fume, hydrotalcite, cement clinker, and calcium sulfate-based expansive agents, have also demonstrated potential for modulating the carbonation performance of geopolymers [34,35]. Overall, research on geopolymer carbonation remains limited compared to OPC systems, particularly given the compositional complexity and variability of geopolymer formulations [36,37]. There is a critical need to elucidate how specific precursor types and ratios influence microstructural evolution and chemical response during carbonation. In particular, the role of MgO in MK-FA blends, and the interaction between MgO and the aluminosilicate gel phases under carbonation conditions, require in-depth investigation.

To address these research gaps, this study systematically investigates the synergistic effects of MgO dosage and metakaolin/fly ash (MK/FA) ratio on the carbonation resistance of MF-based geopolymers. Accelerated carbonation tests were employed to investigate the effects of MgO content and the MK/FA ratio on compressive strength, carbonation depth, and pore solution alkalinity. Subsequently, mercury intrusion porosimetry (MIP), X-ray diffraction (XRD), and Fourier transform infrared spectroscopy (FT-IR) were used to analyze pore structure and phase composition evolution pre- and post-carbonation. This work aims to provide new insights into the formulation design of MgO-modified MF-geopolymers with enhanced carbonation durability, highlighting the unique role of MgO beyond conventional stabilization mechanisms.

2. Experimental Programs

2.1. Raw Materials

In this study, metakaolin (MK) and fly ash (FA) were employed as silica-aluminate precursors to prepare metakaolin–fly ash-based (MF-based) geopolymers. The chemical compositions of MK and FA were characterized using X-ray fluorescence (XRF) spectrometry, with detailed mass fractions of constituent oxides provided in Table 1. Magnesium oxide (MgO) was used as a carbonization modification material, which was a white powder with a purity of over 99% (by mass). In addition, the specific surface area of fly ash metakaolin was 420 m^2/kg and 1200 m^2/kg, respectively.

Table 1. Chemical composition of metakaolin and fly ash (%).

Material	SiO_2	Al_2O_3	CaO	Fe_2O_3	MgO	K_2O	Na_2O	SO_3	TiO_2	Others
Metakaolin	53.80	43.20	-	1.10	0.82	0.45	0.18	-	-	0.45
Fly ash	57.96	31.14	3.02	3.86	0.52	2.03	-	0.64	0.52	0.31

The alkaline activator consisted of anhydrous sodium silicate powder (59.6 wt% SiO_2, 21.6 wt% Na_2O), sodium hydroxide flakes, and deionized water. The sodium silicate solution had an initial modulus (M_s = SiO_2/Na_2O) of 2.85, which was adjusted to a target modulus of 1.2 by adding sodium hydroxide flakes. The final activator contained 10% alkaline content (Na_2O by mass).

2.2. Experimental Design and Specimen Preparation

To address the trade-off between the high reactivity of metakaolin (MK) and the lower reactivity, yet beneficial compositional variability of fly ash (FA), MK was employed as the main precursor, with FA used to partially replace MK at substitution levels ranging from 0 to 40 wt%, thus forming metakaolin–fly ash (MF-based) geopolymer. Additionally, MgO (4–8 wt% of total solids) was incorporated to modify the geopolymer matrix, which was a strategy aimed specifically at enhancing carbonation resistance through microstructural refinement and chemical stabilization. This approach not only differentiated our work from previous studies focused primarily on slag- or fly ash-only systems but also allowed us to evaluate the synergistic effects of MgO with varying MK/FA ratios, which remained largely unexplored in the existing literature. All mixtures were prepared with a constant water-to-binder ratio of 0.63. Detailed mix proportions and experimental parameters of MF-based geopolymers are summarized in Table 2.

Table 2. Design parameters and mix proportions of MF-based geopolymer pastes at various MgO contents.

Serial Number	MgO Content (%)	MK (%)	FA (%)	Alkaline Content (%)	Water/Binder Ratio
M10F0-0%		100	0		
M8F2-0%	0	80	20	10%	0.63
M6F4-0%		60	40		
M10F0-4%		100	0		
M8F2-4%	4	80	20	10%	0.63
M6F4-4%		60	40		
M10F0-8%		100	0		
M8F2-8%	8	80	20	10%	0.63
M6F4-8%		60	40		

Cubic paste specimens (50 × 50 × 50 mm^3) were cast by weighing FA and MK according to the mix proportions in Table 2. To ensure statistical reliability, three replicates were prepared for each testing condition, with results presented as average values. The raw materials were combined with a pre-hydrated alkaline activator (aged for 24 h) and mixed in a planetary mixer. The mixture underwent low-speed mixing for 3 minutes, followed by high-speed mixing for an additional 3 minutes to ensure homogeneity. The preparation process of MF-based geopolymer pastes is plotted in Figure 1. The slurry was then poured into molds and cured in a chamber at 22 °C and 95% relative humidity (RH) for 24 h. After demolding, the specimens were sealed with plastic film and returned to the conditioning chamber (20 ± 2 °C, 95 ± 2% RH) until 28 days of age. Accelerated carbonation tests were initiated during this period.

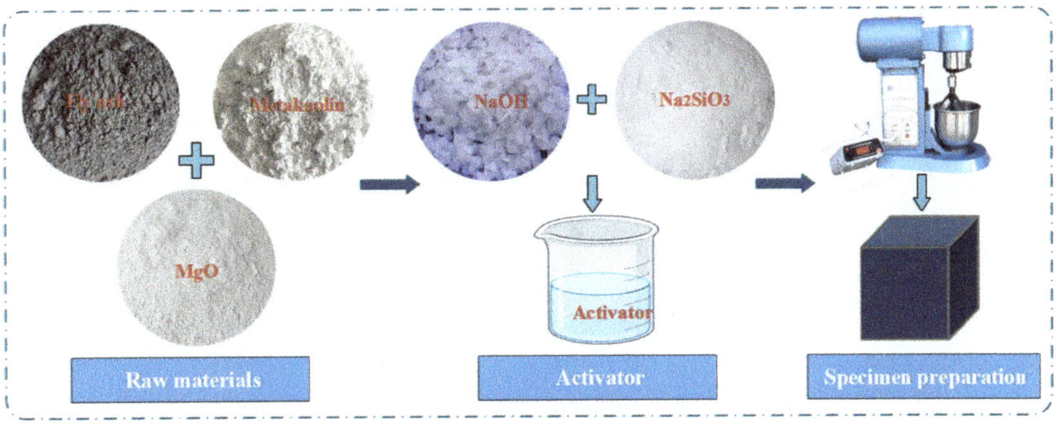

Figure 1. Raw materials and sample preparation of MF-based geopolymer pastes.

2.3. Accelerated Carbonation Procedure

Following the Chinese Standard (GB/T 50082-2024), specimens were preconditioned at 20 °C and 70% RH for 48 h before accelerated carbonation testing. One surface of each specimen was exposed to carbonation, while the remaining surfaces were sealed with paraffin wax. The carbonation chamber maintained a CO_2 concentration of 20 ± 2%, a temperature of 25 ± 3 °C, and an RH of 70 ± 5%. Specimens were subjected to carbonation for durations of 0, 1, 3, 7, 14, 21, and 28 days, with subsequent analyses performed at each interval.

2.4. Testing Methods

2.4.1. Compressive Strength

A hydraulic universal testing machine was used to evaluate the compressive strength of uncarbonated specimens and those that had been carbonated for 28 days. Three replicates per group were tested, and the average value was reported as the compressive strength for that group.

2.4.2. Carbonation Depth

Carbonated specimens were split along their central axis using a dry-cutting saw. A 1% phenolphthalein ethanol solution was sprayed onto the freshly exposed surface. The carbonation front (non-pink region) was measured at ten equidistant points using a vernier caliper (0.1 mm precision). The average distance from the exposed surface was recorded as the carbonation depth.

2.4.3. Material Alkalinity

After measuring carbonation depth, the remaining half of each specimen was sectioned into five depth intervals: 0–10 mm, 10–20 mm, 20–30 mm, 30–40 mm, and 40–50 mm from the exposed surface. Each segment was dried at 50 °C to constant mass, ground, and sieved through a 200-mesh sieve (75 μm aperture). The pH of the pore solution was determined via solid–liquid extraction. Powdered samples were mixed with deionized water (1:3 mass ratio), soaked for 1 hour, and the supernatant pH was measured using a Metrohm 818 Titrino Plus potentiometric titrator (Herisau, Switzerland; ±0.01 accuracy).

2.4.4. MIP, XRD, and FTIR Measurements

Fragments from the 0–10 mm depth of 28-day cured and carbonated specimens were immersed in anhydrous ethanol to arrest hydration, dried, and subjected to mercury intrusion porosimetry (MIP) using a US Conta PoreMaster 33 series aperture analyzer. The XRD tests were carried out using a Rigaku Ultima IV series X-ray diffractometer from Japan with the scanning range set at 5–70° to analyze the powder in the 0–10 mm fraction of the 28-day cured age and 28-day carbonized age samples that had passed through a 200 mesh standard sieve.

Additionally, FTIR testing was performed using a Japanese Shimadzu IRTracer-100 series Fourier transform infrared spectrometer by mixing the powder with KBr and pressing it into a tablet for analysis within a scanning range of 400 to 4000 cm^{-1}, with a resolution of 4 cm^{-1}.

3. Results and Discussion

3.1. Compressive Strength

Accelerated carbonization tests were conducted on MF-based geopolymers with varying MgO contents, and the compressive strength was measured before and 28 days after carbonization. As plotted in Figure 2, the compressive strength of MF-based geopolymer pastes decreased progressively with increasing fly ash content. Specifically, in the MF-based geopolymer, increasing fly ash content from 0% to 40% reduced compressive strength from 37.6 MPa to 33.2 MPa, which reported a decrease of approximately 11.7%. However, the incorporation of MgO significantly enhanced the compressive strength of geopolymer pastes, and this strengthening effect became more pronounced as the fly ash content increased. For the pure metakaolin-based geopolymer system, adding 4% and 8% MgO increased compressive strength by 1.9% and 4.8%, respectively. In MF-based systems with 20% fly ash content, 4% and 8% MgO additions increased compressive strength by 8.6% and 13.9%, respectively. When the fly ash content increased to 40%, the compressive strength increased by 6.0% and 31.3% after adding 4% and 8% MgO, respectively. This amplified strengthening effect at higher fly ash levels is attributed to highly reactive MgO promoting fly ash hydration, while $Mg(OH)_2$ crystals formed during MgO hydration fill internal pores and refine the microstructure, thereby enhancing macroscopic mechanical properties [31,38].

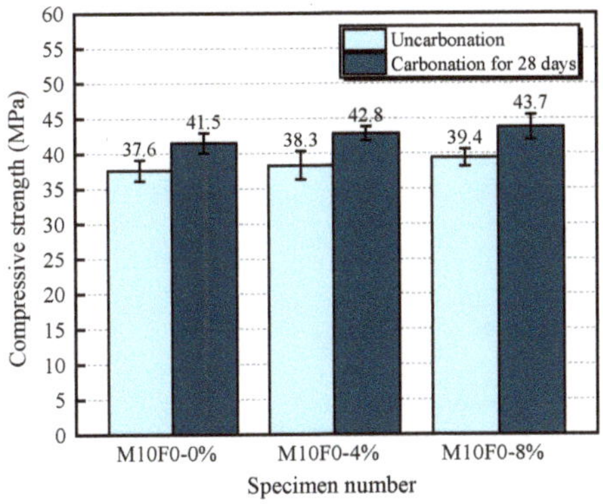

(a) M10F0

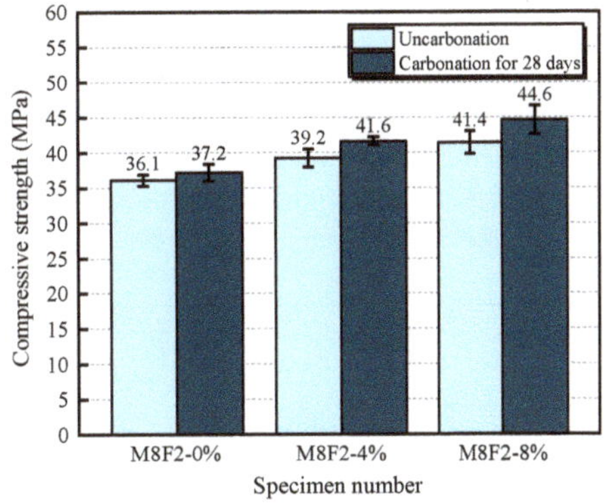

(b) M8F2

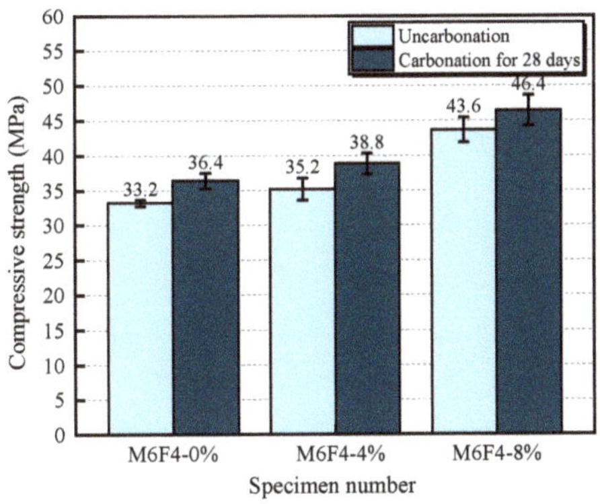

(c) M6F4

Figure 2. The compressive strength of MF-based geopolymer pastes at various MgO contents after carbonation.

Comparing the compressive strength of geopolymer systems with different MgO content before and after carbonization, it was found that the strength of the samples increased after carbonization. For pure metakaolin-based geopolymers, the compressive strength of samples containing 0%, 4%, and 8% MgO increased by 10.4%, 11.7%, and 10.9%, respectively, compared to the uncarbonized samples after carbonization. In the MF geopolymer system with a fly ash content of 20%, the compressive strength of the samples without MgO content and those with 4% and 8% MgO increased by 3.0%, 6.1%, and 7.7%, respectively, after carbonization. When the fly ash content was increased to 40%, the corresponding increases were 9.6%, 10.2%, and 6.4%, respectively. The improvement in post-carbonation strength may be related to the effective sequestration of CO_2 by the hydration reaction-generated hydrotalcite-like phase in the slurry, while the formation of amorphous magnesium-containing hydrates (M-S-H gel) further optimized the microstructure of the geopolymer, thereby enhancing its macroscopic mechanical properties [19].

3.2. Carbonation Depth

Figure 3 illustrates the impact of MgO content on the carbonization depth of MF-based geopolymer pastes. The results revealed that as the carbonization age increased, the carbonization depth of all geopolymer pastes demonstrated an upward trend. However, at the same carbonization age, the specimens containing MgO exhibited significantly lower carbonization depths compared to those specimens without MgO. This enhanced resistance was attributed to two mechanisms: (1) the formation of magnesium-containing layered double hydroxides within the paste, which possess high CO_2 adsorption capacity, and (2) the direct reaction of unreacted MgO particles with CO_2 [36,39].

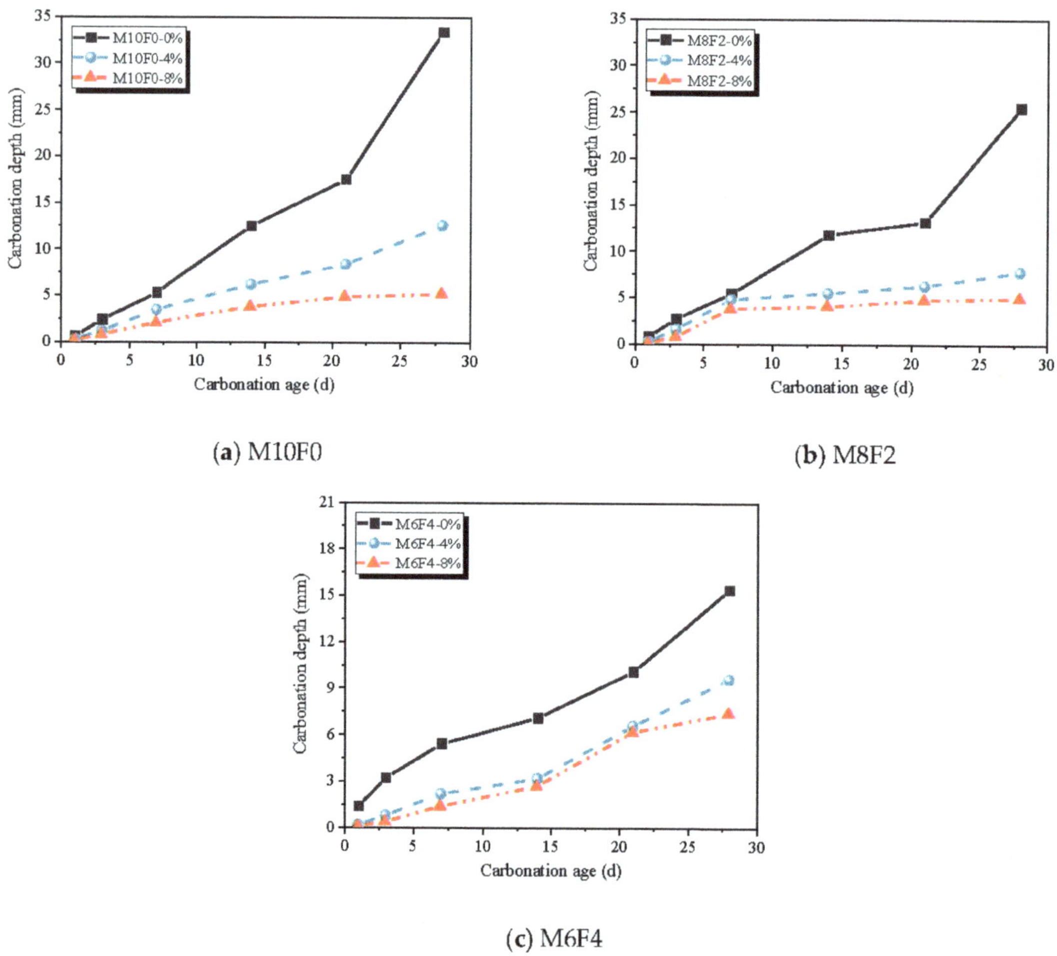

Figure 3. Carbonation depth of MF-based geopolymer pastes at various MgO contents.

In the case of pure metakaolin-based geopolymer pastes with MgO contents of 0%, 4%, and 8%, the carbonization depths of the M10F0-0%, M10F0-4%, and M10F0-8% specimens after 28 days of carbonization were 33.4 mm, 12.6 mm, and 5.2 mm, respectively. In comparison to the geopolymer samples without MgO (M10F0-0%), the carbonation depths of the geopolymer pastes with 4% and 8% MgO additions decreased by 62.3% and 84.4%, respectively. For MF-based geopolymer pastes with 20% fly ash content, the carbonization depths of the M10F0-0%, M10F0-4%, and M10F0-8% specimens after 28 days of carbonization were 25.5 mm, 7.8 mm, and 5.0 mm, respectively, when the MgO content was 0%, 4%, and 8%. Compared to the geopolymers without MgO (M8F2-0%), the carbonation depths of the MF-based geopolymer pastes with 4% and 8% MgO additions were reduced by 69.4% and 80.4%, respectively. For MF-based geopolymer pastes with 40% fly ash content, the carbonization depths of the M10F0-0%, M10F0-4%, and M10F0-8% specimens after 28 days of carbonization were 15.4 mm, 9.6 mm, and 7.4 mm, respectively, when the MgO content was 0%, 4%, and 8%. When compared to the control group without MgO (M6F4-0%), the carbonation depths of the specimens with 4% and 8% MgO additions decreased by 37.7% and 51.9%, respectively. Furthermore, as the fly ash content increased, the carbonization depth of the MF-based geopolymer pastes decreased. Specifically, increasing the fly ash content from 0% to 40% reduced the carbonation depth of MF-based geopolymer pastes from 33.4 mm to 15.4 mm, as demonstrated in Figure 3.

3.3. pH Value

Alkalinity testing was performed on powder sampled from the 0–10 mm depth section extending inward from the exposed surface of the specimens. Figure 4 presents the pH values measured for each group of MF-based geopolymer pastes at different carbonation ages. A gradual decline in pH value was observed for all MF-based geopolymer pastes as the carbonation age increased. This is because as the geopolymer sample continues to be exposed to the CO_2 environment, the concentration gradient between the interior and exterior of the slurry causes CO_2 to continuously diffuse inward, reacting with the hydration products of geopolymers, ultimately leading to a gradual decrease in the alkalinity of the pastes [13,15]. Notably, the pore solution pH decreased significantly during the first 7 days of carbonation, after which the rate of decrease slowed and gradually stabilized. This stabilization primarily occurs because ions within the geopolymer pore solution reached an equilibrium state after 28 days of carbonation. Consequently, the final pore solution pH demonstrated no significant difference between the metakaolin-based geopolymer paste and the MF-based geopolymer pastes containing either 4% or 8% MgO.

During the initial carbonation stage, pH values for all geopolymer pastes ranged from 11.92 to 12.31. After 28 days of carbonation, the alkalinity of unmodified MF-based geopolymer paste fell below 10.5. This decrease in pH value was directly proportional to the depth of carbonation, as indicated in Figure 3. The incorporation of MgO significantly reduced the rate of pH decline. This was primarily attributed to MgO effectively delaying the ingress of external CO_2, thereby slowing the carbonation rate. For the pure metakaolin-based geopolymer paste, when the MgO content was 0%, 4%, and 8%, the pH values after 28 days of carbonation were 10.36, 10.81, and 10.76, respectively, which decreased by 1.65, 1.11, and 1.22 compared to those of specimens before carbonation. For the MF-based geopolymer paste with 20% fly ash content, when the MgO content was 0%, 4%, and 8%, the pH values after 28 days of carbonation were 9.78, 11.00, and 10.93, respectively, which decreased by 2.40, 0.96, and 1.13 compared to those of specimens before carbonation. For the MF-based geopolymer paste with 40% fly ash content, when the MgO content was 0%, 4%, and 8%, the pH values after 28 days of carbonation were 10.49, 11.32, and 11.33, respectively, which decreased by 1.74, 0.93, and 0.98 compared to those of specimens before carbonation. These results demonstrated that MgO modification maintained the pore solution pH around 11 after 28 days of carbonation. Additionally, increasing the fly ash content reduced the initial pH of the uncarbonated paste. When the fly ash content increased from 0% to 40% in the MF-based geopolymer pastes, the pH value exhibited a slight decreasing trend (approximately 0.2).

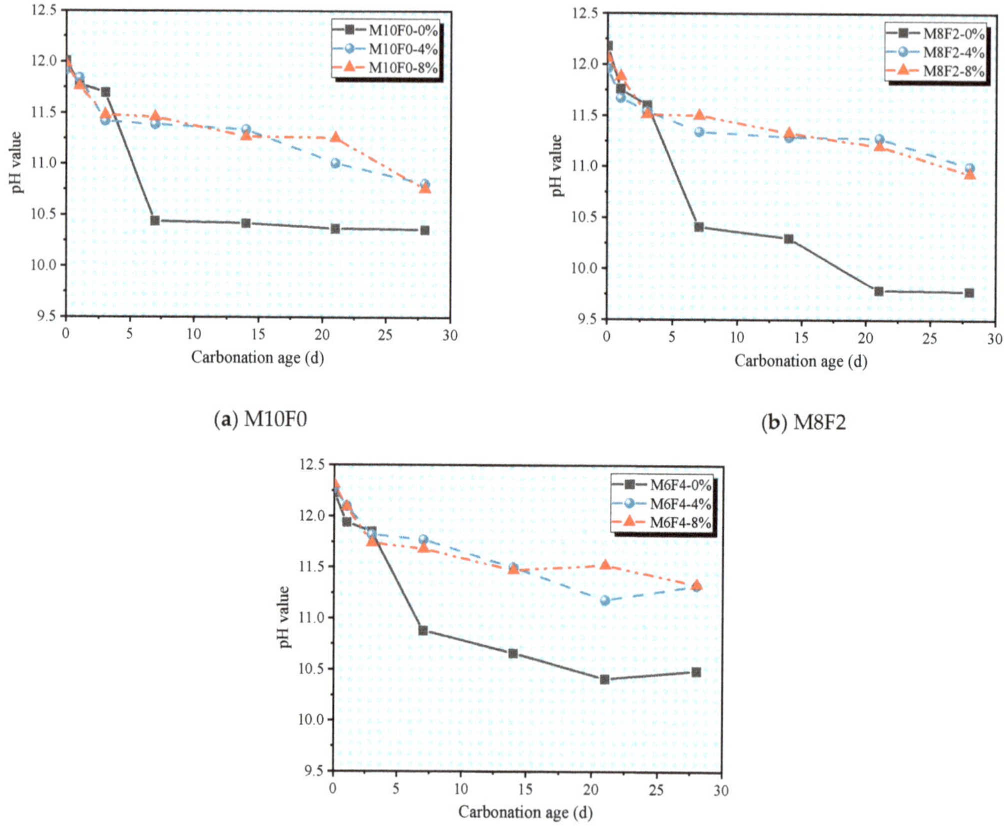

(**a**) M10F0 (**b**) M8F2 (**c**) M6F4

Figure 4. The alkalinity of MF-based geopolymer pastes with various MgO contents.

3.4. Pore Structure

Figure 5 illustrates the changes in pore size distribution, porosity, and the most probable pore size of the M8F2-0% and M8F2-8% specimens before and after carbonation. Pore sizes in cementitious materials were typically categorized into four types: gel micropores (<10 nm), mesopores (10–50 nm), capillary pores (50–100 nm), and macropores (100–1000 nm) [1]. As depicted in Figure 5a, the pore size distribution of the M8F2-0% control group exhibited minimal variation before and after carbonation, with pore concentrations predominantly in the mesopore (10–50 nm) range. The changes in capillary and macropores were negligible. The absence of MgO in this group resulted in changes in porosity, primarily attributed to the transformation of $Ca(OH)_2$ and CO_2 infiltration during the carbonation process. The limited alteration in pore size may stem from the relatively loose initial microstructure of the control group, which meant that CO_2 invasion and $Ca(OH)_2$ transformation had a constrained impact on pore size distribution.

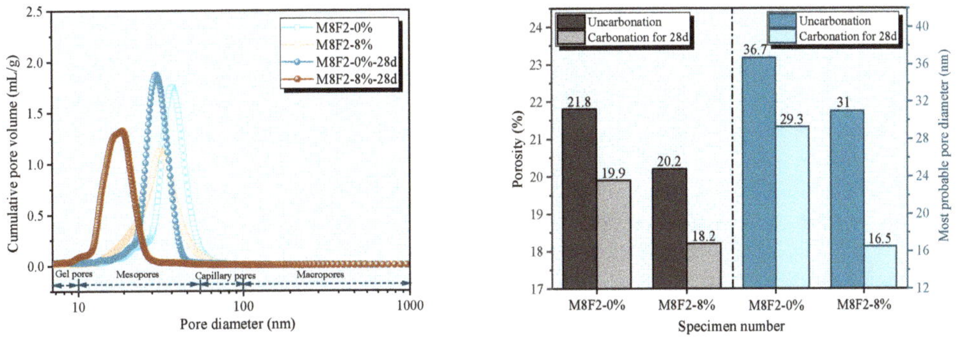

(**a**) Pore size distribution (**b**) Porosity

Figure 5. Pore size distribution and porosity of MF-based geopolymer pastes modified with MgO content.

In contrast, after the introduction of MgO, the pore size distribution curve for the M8F2-8% specimens shifted leftward compared to the control group (M8F2-0%) prior to carbonation. This shift indicated a decrease in both capillary pores (50–100 nm) and macropores (>100 nm), suggesting an enhancement in the pore structure. This improvement manifested in the inclusion of MgO, which optimized the microstructure of the paste, ultimately reducing porosity and the most probable pore size. This enhanced porosity contributed to the enhancement of the compressive strength of the pastes, as demonstrated in Figure 2. The underlying mechanism involved MgO accelerating the hydration reaction and facilitating the formation of finer $Mg(OH)_2$ crystals within the highly alkaline pore solution. These finer crystals were better dispersed throughout the paste, thereby enhancing the microstructure [40]. After 28 days of carbonation, the pore size distribution curve for the M8F2-8% specimens shifted significantly to the left, indicating a marked reduction in pore diameter.

Moreover, both the total porosity and the most probable pore diameter for the M8F2-0% and M8F2-8% specimens decreased following carbonation. Specifically, after 28 days, the porosity of the M8F2-0% and M8F2-8% specimens decreased from 21.8% and 20.2% to 19.9% and 18.2%, respectively. Concurrently, their most probable pore diameters diminished from 36.7 nm and 31.0 nm to 29.3 nm and 16.5 nm, respectively. This refinement of pores post-carbonation can be attributed to the continued participation of some $Mg(OH)_2$ in the reaction within the carbonation environment, leading to the formation of expansive magnesium silicate hydrate (M-S-H) gel or a similar hydrotalcite phase [36]. Additionally, unreacted MgO particles can directly carbonize to produce magnesium carbonate ($MgCO_3$) and hydrated magnesium carbonate. These carbonized products exhibit excellent bonding and filling characteristics, effectively refining the pores and reducing the overall porosity of the paste. Consequently, the macroscopic mechanical properties of the geopolymer incorporating MgO were enhanced after carbonation, aligning with the results of the compressive strength, as illustrated in Figure 5.

3.5. XRD Analysis

Figure 6 presents the XRD patterns of the M8F2-0% and M8F2-8% specimens, both before carbonization and 28 days post-carbonization. Within the 2θ range of 15 to 35°, the MgO-modified geopolymers exhibited several characteristic peaks indicative of an amorphous structure. The detected crystalline phases included quartz, mullite, and minor amounts of hydrotalcite and akermanite, along with (N,C,m)-A-S-H gel as the primary hydration products [26,41]. Notably, the hydration products from the MF-based geopolymer pastes did not contain calcium hydroxide or C-S-H gel, which were essential calcium reserves for counteracting carbonization. In contrast, the introduction of MgO to the MF-based geopolymer pastes resulted in a reduction in the intensity of the corresponding peaks, suggesting that MgO significantly enhanced the hydration reaction within the MF-based geopolymer pastes. This enhancement led to an increased number of amorphous gel phases, as reflected by the changes in the characteristic peaks. The phase composition of the M8F2-0% specimen remained largely unchanged before and after carbonization, though a small quantity of sodium salt crystals was observed following carbonization. In the hydration products of the MF-based geopolymer pastes with MgO addition, a distinct periclase diffraction peak was evident, indicating that MgO has not completely reacted. After carbonation, the MgO diffraction peak disappeared, indicating its reaction with CO_2. Although hydrotalcite and akermanite were detected, their origin may be related to the raw materials rather than resulting solely from hydration. It should be noted that the expected carbonation product, such as magnesite ($MgCO_3$), was not clearly detected by XRD, possibly due to its low crystallinity or content. This suggested direct carbonation of MgO may have formed poorly crystalline or amorphous phases.

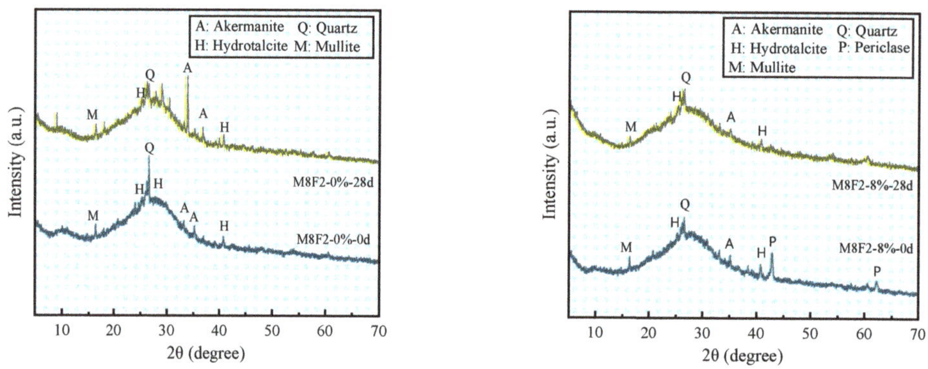

(a) M8F2

(b) M8F2-8%

Figure 6. XRD patterns of MF-based geopolymer pastes modified with MgO before and after carbonation.

3.6. FT-IR Analysis

Figure 7 presents the FT-IR spectra of MM8F2-0% and M8F2-8% specimens before carbonation and after 28 days of carbonation. Analysis of specific functional groups within the hydration and carbonation products enables the assessment of qualitative material changes induced by carbonation. Overall, the phase composition of the MF-based geopolymer pastes remained largely unchanged following carbonation. Consequently, the functional groups associated with bound water, carbonates, and the gel structure exhibited no significant alterations. However, carbonation increased the polymerization degree of the silicate gel. Analysis of the pre-carbonation FT-IR spectra revealed absorption peaks at 3371 cm^{-1} and 3403 cm^{-1} for M8F2-0% and M8F2-8%, respectively, originating from O-H stretching vibrations of water molecules [42]. After 28 days of carbonation, these peaks shifted to 3368 cm^{-1} and 3367 cm^{-1}. Peaks observed at 1654 cm^{-1} and 1647 cm^{-1} were attributed to O-H bending vibrations of bound water within the gel. Absorption peaks at 1458 cm^{-1} and 1363 cm^{-1} arose from asymmetric stretching vibrations of carbonate species, likely formed from unreacted residual NaOH present in the activator [6]. Following carbonation, these peaks shifted to higher wavenumbers (1503 cm^{-1} for M8F2-0% and

1385 cm^{-1} for M8F2-8%), consistent with substantial OH^- consumption to counteract CO_2 ingress and generate carbonates.

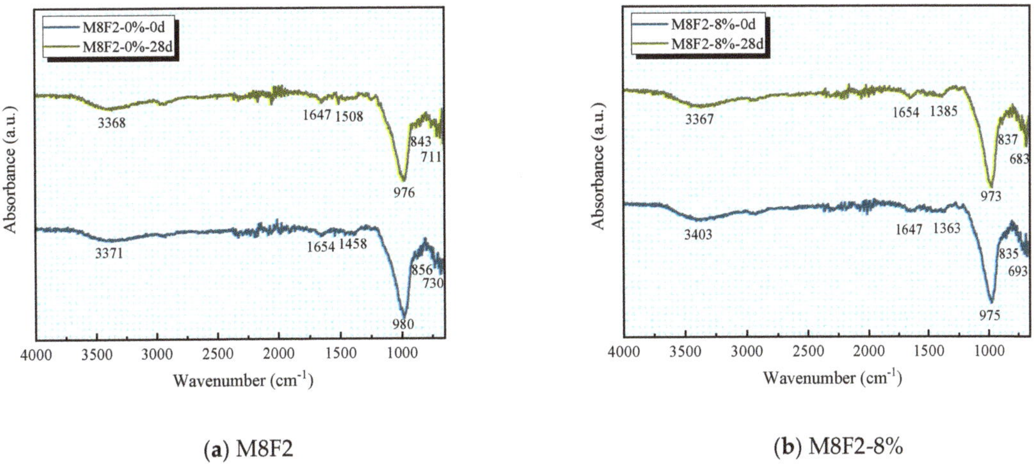

Figure 7. FT-IR spectra of MF-based geopolymer pastes modified with MgO before and after carbonation.

Additionally, distinct peaks at 980 cm^{-1} (M8F2-0%) and 975 cm^{-1} (M8F2-8%) before carbonation correspond to asymmetric stretching vibrations of Si-O-Si and Si-O-Al bonds in the gel [43,44]. After 28 days of carbonation, these peaks shifted to 976 cm^{-1} and 973 cm^{-1}, respectively, indicating a movement to lower wavenumbers. Incorporation of MgO induced a shift towards lower wavenumbers for these peaks in both pre- and post-carbonation specimens. The wavenumber of this peak served as an indicator of the gel's polymerization state and degree within the matrix. MgO addition promoted the replacement of Al^{3+} by Si^{4+} during reaction, transforming aluminum-rich geopolymeric gel into a more stable silicon-rich structure [38,45]. This transformation exhibited lower porosity and higher strength in the geopolymer pastes, as illustrated in Figures 2 and 5. This observation indicated that MgO not only enhanced the hydration reaction but also facilitated the formation of structurally more stable gels, thereby contributing to improved macroscopic material strength. Furthermore, absorption peaks at 856 cm^{-1} (M8F2-0%) and 835 cm^{-1} (M8F2-8%) were potentially associated with symmetric Si-O-Si stretching vibrations. Peaks occurring between 500 cm^{-1} and 800 cm^{-1} were attributed to in-plane bending vibrations of Si-O or Al-O bonds within the constituent SiO_4 and AlO_4 tetrahedral structural units [1,46].

4. Conclusions

This study investigates the impact of varying MgO and fly ash content on the carbonation performance of MF-based geopolymer pastes through accelerated carbonation tests. The carbonation mechanisms of MF-based geopolymer pastes at different ages were elucidated through XRD, FTIR, and MIP. The key findings are as follows:

(1) The compressive strength of MF-based geopolymer pastes gradually decreased with increasing fly ash content. Specifically, when the fly ash content rose from 0% to 40%, the compressive strength decreased by 11.7%. Conversely, the inclusion of MgO enhanced the compressive strength of the MF-based geopolymer pastes, with a more pronounced improvement observed at higher fly ash contents. For instance, at an 8% MgO content, the compressive strength of the M10F0-8%, M8F2-8%, and M6F4-8% specimens increased by 4.8%, 14.7%, and 31.3%, respectively, compared to the control group without

MgO. Furthermore, carbonation led to a moderate strength gain in all specimens, with improvements ranging from 3% to 11%.

(2) As the carbonation age progressed, the carbonation depth of geopolymer pastes increased gradually, while the pH value of the pore solution declined. The incorporation of MgO and fly ash effectively mitigated both the progression of carbonation depth and the drop in pH value. For the MF-based geopolymer pastes with 20% fly ash content, the carbonation depths after 28 days were measured at 25.5 mm, 7.8 mm, and 5.0 mm for MgO contents of 0%, 4%, and 8%, respectively. In comparison to the geopolymer specimens without MgO (M8F2-0%), the additions of 4% and 8% MgO reduced the carbonation depth by 69.4% and 80.4%, respectively. After 28 days of carbonation, the pH values for M8F2-0%, M8F2-4%, and M8F2-8% were 9.78, 11.00, and 10.93, representing reductions of 2.40, 0.96, and 1.13 from their initial pre-carbonation values.

(3) The inclusion of MgO reduced the median pore diameter, total porosity, and most probable pore size. This effect was attributed to MgO accelerating the hydration reaction, which promoted the formation of finer $Mg(OH)_2$ crystals in the highly alkaline pore solution. These crystals contribute to a more refined and homogeneous microstructure, thereby enhancing compressive strength. Furthermore, carbonation for 28 days also reduced both the porosity and the most probable pore size. Specifically, the porosity of the M8F2-0% and M8F2-8% samples decreased from 21.8% and 20.2% to 19.9% and 18.2%, respectively, while their most probable pore sizes reduced from 36.7 nm and 31.0 nm to 29.3 nm and 16.5 nm, respectively.

(4) The phase composition of the MF-based geopolymer pastes remained largely unchanged before and after carbonation, with the primary hydration products identified as N-A-S-H gel, mullite, hydrotalcite, and akermanite. The introduction of MgO reduced the intensity of the characteristic peaks for these phases, indicating that MgO significantly promoted the hydration reactions within the geopolymer system.

Author Contributions: Methodology, S.L.; Formal analysis, D.J.; Investigation, S.L.; Data curation, S.L.; Writing—original draft, S.L. and D.J.; Writing—review & editing, S.L. and D.J.; Project administration, D.J. All authors have read and agreed to the published version of the manuscript.

Funding: This research received no external funding.

Data Availability Statement: The original contributions presented in this study are included in the article material. Further inquiries can be directed to the corresponding author.

Acknowledgments: This work was supported by the Natural Science Foundation of Hunan Province, China (Grant No. 2024JJ9067).

Conflicts of Interest: The authors declare no conflict of interest.

References

1. Fu, Q.; Xu, W.; Zhao, X.; Bu, M.; Yuan, Q.; Niu, D. The microstructure and durability of fly ash-based geopolymer concrete: A review. *Ceram. Int.* **2021**, *47*, 29550–29566. [CrossRef]
2. Zhang, P.; Wang, K.; Li, Q.; Wang, J.; Ling, Y. Fabrication and engineering properties of concretes based on geopolymers/alkali-activated binders—A review. *J. Clean. Prod.* **2020**, *258*, 120896. [CrossRef]
3. Yang, G.; Zhao, J.; Wang, Y. Durability properties of sustainable alkali-activated cementitious materials as marine engineering material: A review. *Mater. Today Sustain.* **2022**, *17*, 100099. [CrossRef]
4. Skariah Thomas, B.; Yang, J.; Bahurudeen, A.; Chinnu, S.N.; Abdalla, J.A.; Hawileh, R.A. Geopolymer concrete incorporating recycled aggregates: A comprehensive review. *Clean. Mater.* **2022**, *3*, 100056. [CrossRef]
5. Kumar, B.G.; Selvam, R.; Govindaraj, V. Durability assessment of blended slag-fly ash geopolymer concrete in field conditions after 5 years. *J. Sustain. Cem.-Based Mater.* **2024**, *13*, 1770–1781. [CrossRef]
6. Bernal, S.A.; Provis, J.L. Durability of alkali- activated materials: Progress and perspectives. *J. Am. Ceram. Soc.* **2014**, *97*, 997–1008. [CrossRef]

7. Zhang, B. Durability of low-carbon geopolymer concrete: A critical review. *Sustain. Mater. Technol.* **2024**, *40*, e00882. [CrossRef]
8. Zhang, B.; Wang, Q.; Zhu, H.; Peng, H. Behavior of BFRP-confined geopolymer-based coral aggregate concrete columns under axial compression: Effects of specimen sizes. *J. Build. Eng.* **2024**, *98*, 111106. [CrossRef]
9. Mundra, S.; Criado, M.; Bernal, S.A.; Provis, J.L. Chloride-induced corrosion of steel rebars in simulated pore solutions of alkali-activated concretes. *Cem. Concr. Res.* **2017**, *100*, 385–397. [CrossRef]
10. Badar, M.S.; Kupwade-Patil, K.; Bernal, S.A.; Provis, J.L.; Allouche, E.N. Corrosion of steel bars induced by accelerated carbonation in low and high calcium fly ash geopolymer concretes. *Constr. Build. Mater.* **2014**, *61*, 79–89. [CrossRef]
11. Pasupathy, K.; Sanjayan, J.; Rajeev, P. Evaluation of alkalinity changes and carbonation of geopolymer concrete exposed to wetting and drying. *J. Build. Eng.* **2021**, *35*, 102029. [CrossRef]
12. Hamada, H.M.; Al-Attar, A.; Beddu, S.; Askar, M.K.; Yousif, S.T.; Majdi, A. Impact of rice husk ash on geopolymer concrete: A literature review and future directions. *Case Stud. Constr. Mater.* **2025**, *22*, e04476. [CrossRef]
13. Stefanoni, M.; Angst, U.; Elsener, B. Corrosion rate of carbon steel in carbonated concrete—A critical review. *Cem. Concr. Res.* **2018**, *103*, 35–48. [CrossRef]
14. Zhang, C.; Guo, H.; Shi, L.; Hou, X.; Kong, X.; Yu, B. Corrosion resistance of polymer-modified hardened cement paste and phosphoric acid-activated metakaolin geopolymer in carbonic acid solution. *Constr. Build. Mater.* **2024**, *445*, 137950. [CrossRef]
15. Bernal, S.A.; Provis, J.L.; Brice, D.G.; Kilcullen, A.; Duxson, P.; Van Deventer, J.S.J. Accelerated carbonation testing of alkali-activated binders significantly underestimates service life: The role of pore solution chemistry. *Cem. Concr. Res.* **2012**, *42*, 1317–1326. [CrossRef]
16. Zhang, X.; Long, K.; Liu, W.; Li, L.; Long, W.J. Carbonation and chloride ions' penetration of alkali-activated materials: A review. *Molecules* **2020**, *25*, 5074. [CrossRef] [PubMed]
17. Zhang, B.; Zhu, H.; Yang, Z.; Dong, Y.R. BFRP bars reinforced geopolymer-based coral aggregate concrete beams with sustainable and high seawater erosion resistance: Flexural durability, economic, and ecological analysis. *Eng. Struct.* **2025**, *330*, 119910. [CrossRef]
18. Lv, Y.; Qiao, J.; Han, W.; Pan, B.; Jin, X.; Peng, H. Modification effect of $Ca(OH)_2$ on the carbonation resistance of fly ash-metakaolin-based geopolymer. *Materials* **2023**, *16*, 2305. [CrossRef]
19. Yuan, W.B.; Zheng, Y.; Yu, N.T.; Shen, Y.; Hao, X.K. Comparative analysis of time-dependent CO_2 uptake in cement and geopolymer concretes via diffusion-induced carbonation. *Case Stud. Constr. Mater.* **2025**, *22*, e04573. [CrossRef]
20. Zhang, Z.; Bu, Y.; Guo, S.; Lu, C.; Liu, H.; Guo, X. Experimental and thermodynamic analysis of the carbonation of geopolymer under geologic CO_2 sequestration conditions. *Phys. Fluids* **2025**, *37*, 043356. [CrossRef]
21. Cao, J.; Peng, H.; Zhang, B.; Yang, Z.; Pan, H.; Chen, J.; Zhao, X. Basalt fiber-reinforced seawater sea-sand geopolymer composites for offshore engineering constructions: Mechanical performance, microstructure, enhanced mechanisms and sustainability. *Constr. Build. Mater.* **2025**, *493*, 143129. [CrossRef]
22. Abdulhaleem, K.N.; Hamada, H.M.; Osman, A.I.; Yousif, S.T.; Humada, A.M.; Majdi, A. A comprehensive review of sustainable geopolymer concrete using palm oil clinker: Environmental and engineering aspects. *Energy Sci. Eng.* **2025**, *13*, 958–979. [CrossRef]
23. Zhang, B.; Zhu, H.; Cheng, Y.; Huseien, G.F.; Shah, K.W. Shrinkage mechanisms and shrinkage-mitigating strategies of alkali-activated slag composites: A critical review. *Constr. Build. Mater.* **2022**, *318*, 125993. [CrossRef]
24. Li, N.; Farzadnia, N.; Shi, C. Microstructural changes in alkali-activated slag mortars induced by accelerated carbonation. *Cem. Concr. Res.* **2017**, *100*, 214–226. [CrossRef]
25. Abdalqader, A.; Jin, F.; Al-Tabbaa, A. Performance of magnesia-modified sodium carbonate-activated slag/fly ash concrete. *Cem. Concr. Compos.* **2019**, *103*, 160–174. [CrossRef]
26. Zuo, Y.; Nedeljković, M.; Ye, G. Pore solution composition of alkali-activated slag/fly ash pastes. *Cem. Concr. Res.* **2019**, *115*, 230–250. [CrossRef]
27. Robayo-Salazar, R.A.; Aguirre-Guerrero, A.M.; Mejía de Gutiérrez, R. Carbonation-induced corrosion of alkali-activated binary concrete based on natural volcanic pozzolan. *Constr. Build. Mater.* **2020**, *232*, 117189. [CrossRef]
28. Bakharev, T.; Sanjayan, J.G.; Cheng, Y.B. Resistance of alkali-activated slag concrete to carbonation. *Cem. Concr. Res.* **2001**, *31*, 1277–1283. [CrossRef]
29. Zhao, C.; Li, Z.; Peng, S.; Liu, J.; Wu, Q.; Xu, X. State-of-the-art review of geopolymer concrete carbonation: From impact analysis to model establishment. *Case Stud. Constr. Mater.* **2024**, *20*, e03124. [CrossRef]
30. Elhag, A.B.; Selmi, A.; Ahmad, Z.; Ghazouani, N. Enhancing mechanical properties and microstructural integrity of silica fume-based self-compacting geopolymer concrete through municipal solid waste incineration fly ash incorporation. *Mater. Lett.* **2025**, *392*, 138541. [CrossRef]
31. Zhang, M.; Zhang, H.; Tang, H.; Sun, Y.; Bu, C.; Zhang, J. Improving the carbonation resistance of alkali-activated slag mortars by MgO with different reactivity. *Mater. Struct.* **2025**, *58*, 77. [CrossRef]

32. Bernal, S.A.; San Nicolas, R.; Myers, R.J.; Mejía de Gutiérrez, R.; Puertas, F.; Van Deventer, J.S.J. MgO content of slag controls phase evolution and structural changes induced by accelerated carbonation in alkali-activated binders. *Cem. Concr. Res.* **2014**, *57*, 33–43. [CrossRef]
33. Lv, Y.; Pan, B.; Han, W.; Qiao, J.; Zhang, B.; Zhang, J. Carbonation resistance of fly ash-metakaolin-based geopolymer pastes regulated by slag and carbide slag. *J. Sustain. Cem.-Based Mater.* **2024**, *13*, 1538–1551. [CrossRef]
34. Zhang, P.; Gao, Z.; Wang, J.; Guo, J.; Hu, S.; Ling, Y. Properties of fresh and hardened fly ash/slag based geopolymer concrete: A review. *J. Clean. Prod.* **2020**, *270*, 122389. [CrossRef]
35. Haq, E.U.; Padmanabhan, S.K.; Licciulli, A. In-situ carbonation of alkali activated fly ash geopolymer. *Constr. Build. Mater.* **2014**, *66*, 781–786. [CrossRef]
36. Lamaa, G.; Duarte, A.P.C.; Silva, R.V.; De Brito, J. Carbonation of alkali-activated materials: A review. *Materials* **2023**, *16*, 3086. [CrossRef]
37. Ye, H.; Cai, R.; Tian, Z. Natural carbonation-induced phase and molecular evolution of alkali-activated slag: Effect of activator composition and curing temperature. *Constr. Build. Mater.* **2020**, *248*, 118726. [CrossRef]
38. Zhang, Y.; Wang, W.; Lv, Q. The influence of reactive MgO on the hydration and carbonation performance of slag-rich cement system. *J. Build. Eng.* **2023**, *77*, 107477. [CrossRef]
39. Podnar, T.M.; Knez, Z.; Kravanja, G. Enhancing strength and CO_2 uptake in lignite-based fly ash geopolymer mortar through supercritical carbonation. *J. Supercrit. Fluid.* **2025**, *225*, 106695. [CrossRef]
40. Hay, R.; Peng, B.; Celik, K. Filler effects of $CaCO_3$ polymorphs derived from limestone and seashell on hydration and carbonation of reactive magnesium oxide (MgO) cement (RMC). *Cem. Concr. Res.* **2023**, *164*, 107040. [CrossRef]
41. Provis, J.L.; Fernández-Jiménez, A.; Kamseu, E.; Leonelli, C.; Palomo, A. Binder chemistry—Low-calcium alkali-activated materials. In *Alkali Activated Materials*; Springer: Berlin/Heidelberg, Germany, 2014; Volume 1, pp. 125–144. [CrossRef]
42. Salim, M.U.; Moro, C. Microstructural insights of geopolymer mortar containing cenosphere: Effects on fresh properties and durability. *Mater. Struct.* **2025**, *58*, 101. [CrossRef]
43. Elfadaly, E.; Othman, A.M.; Aly, M.H.; Elgarhy, W.A.; Abdellatief, M. Assessing performance and environmental benefits of high-performance geopolymer mortar incorporating pumice and rice straw ash. *Sustain. Chem. Pharm.* **2025**, *44*, 101918. [CrossRef]
44. Xue, C.; Sirivivatnanon, V.; Nezhad, A.; Zhao, Q. Comparisons of alkali-activated binder concrete (ABC) with OPC concrete-A review. *Cem. Concr. Compos.* **2023**, *135*, 104851. [CrossRef]
45. Lee, N.K.; Koh, K.T.; Kim, M.O.; An, G.H.; Ryu, G.S. Physicochemical changes caused by reactive MgO in alkali-activated fly ash/slag blends under accelerated carbonation. *Ceram. Int.* **2017**, *43*, 12490–12496. [CrossRef]
46. Bu, L.; Sun, R.; Guan, Y.; Fang, C.; Ge, Z.; Ran, Y. Effect of white mud on carbonation resistance of alkali activated slag. *Constr. Build. Mater.* **2024**, *438*, 137207. [CrossRef]

Disclaimer/Publisher's Note: The statements, opinions and data contained in all publications are solely those of the individual author(s) and contributor(s) and not of MDPI and/or the editor(s). MDPI and/or the editor(s) disclaim responsibility for any injury to people or property resulting from any ideas, methods, instructions or products referred to in the content.

Journal of
Composites Science

MDPI

Article

Analytical and Numerical Investigation of Vibration Characteristics in Shear-Deformable FGM Beams

Murat Çelik [1], Erol Demirkan [2] and Ahmet Feyzi Yıldırım [2,*]

1 Instituto de Ciencia y Tecnología de Polímeros ICTP, Consejo Superior de Investigaciones Científicas (CSIC), C/Juan de la Cierva, 3, 28006 Madrid, Spain
2 Department of Civil Engineering, Istanbul Technical University, 34469 Istanbul, Turkey
* Correspondence: yildirima15@itu.edu.tr

Abstract: In this study, the free vibration characteristics of a functionally graded (FG) shear-deformable Timoshenko beam were investigated both analytically and numerically. The work is notable for its significant contribution to the literature, particularly in addressing analytically challenging problems related to complex FGM structures using advanced computer-aided finite element methods. For the analytical approach, the governing equations and associated boundary conditions were derived using Hamilton's principle of minimum potential energy. These equations were then solved using the Navier solution method to determine the natural frequencies of the beam. In the numerical analysis, a 3D FG beam model was developed in the ABAQUS finite element software (2023, Dassault Systèmes, Providence, RI, USA)using the second-order hexahedral (HEX20/C3D20) and 1D three-node quadratic beam (B32) elements. The material gradation was defined layer-by-layer along the thickness direction in accordance with the rule of mixtures. Modal analysis was subsequently performed to extract the natural frequency values. The results show a high level of agreement between the analytical and numerical solutions. and were consistent with previously published studies in the literature.

Keywords: FGM beam; Timoshenko beam; finite element; composite structures

1. Introduction

Functionally graded materials (FGMs) are advanced engineering materials characterized by a gradual and continuous variation in their constituent composition, which allows controlled changes in mechanical, thermal, or electrical properties compared to conventional homogeneous materials. This unique feature enables FGMs to overcome the limitations of traditional materials, providing the ability to meet different performance requirements in different regions and optimize material usage. In particular, when applied to beams and slender structural elements, FGMs can enhance load-bearing capacity while reducing overall weight [1]. Analytical and numerical studies have shown a strong relationship between the material gradient and the elastic, thermomechanical, and dynamic behavior of beams, highlighting the critical role of material distribution in design processes [2].

Experimental research has also demonstrated that FGM beams can improve resistance to temperature variations, thermal stresses, and critical load conditions. Studies on micro- and nano-scale structures indicate that FGM beams offer more precise performance control than conventional homogeneous materials, enabling designers to achieve desired stiffness, flexibility, and vibration characteristics through gradient optimization. Moreover, recent advanced numerical simulations and modeling studies have provided valuable insights into predicting the behavior of FGM micro-beams carrying attached mass and boundary conditions, offering essential guidance for material and structural design [3].

In engineering, FGM beams offer the opportunity to optimize multiple design criteria such as weight, strength, and durability simultaneously, particularly in aerospace, automotive, biomedical, and micro-electromechanical systems (MEMS/NEMS). The distinguishing feature of FGMs is the continuous spatial variation of properties of two or more materials, especially ceramics and metals. This graded structure provides significant advantages, particularly in terms of heat resistance, the prevention of crack propagation, and reductions in thermal stresses. FGMs greatly minimize the issues arising from interface weaknesses and abrupt material transitions, which are commonly encountered in conventional layered composites [4].

Vibration analysis is one of the focal points of research on FGM beams. The effects of material gradation on natural frequencies and mode shapes have been examined in detail using both classical and advanced theories in cases of free and forced vibration. Alshorbagy et al. [5] studied the free vibration behavior of FGM beams using the finite element method, and revealed the impact of the material gradient index and geometric parameters on frequencies. Similarly, Avcar [6] and Kahya and Turan [7] analyzed changes in natural frequencies under different boundary conditions and various loading scenarios.

The mechanical response of functionally graded structures varies substantially depending on the material gradient function, cross-sectional geometry, and boundary conditions. The bending [8], free and force [9,10] vibrations, thermal [11,12], and buckling [13] behaviors of FGM beams and plates were studied using various solution methods. Pradhan and Chakraverty [14] comprehensively examined the free vibration of Euler and Timoshenko FGM beams using the Rayleigh–Ritz method. Both numerical and experimental studies have shown how material gradation in FGMs manages thermal stresses and prevents crack propagation [15]. Kahya and Turan [7,16] analyzed the vibration and stability behaviors of FGM and sandwich beams using the first-order shear deformation theory and multilayer finite element methods. Nguyen and Nguyen [17] adapted higher-order shear deformation theory to FGM sandwich beams, and achieved significant advantages in static, buckling, and free vibration analyses.

More recently, Echouai et al. [18] presented a finite element model using Euler–Bernoulli beam theory to analyze transverse vibrations in FGM beams with Terfenol-D magnetostrictive layers on a Winkler foundation. It explores the influence of material and structural parameters on natural frequencies and damping, offering insights for improved vibration control and structural design. Benadouda et al. [19] analyzed wave propagation in functionally graded (FG) pinned–pinned beams with defects such as cracks and bidirectional porosity using higher-order shear deformation theory and Hamilton's principle. Jayachandiran et al. [20] conducted both experimental and numerical analyses to examine the free and forced vibration behaviors of functionally graded material (FGM) beams reinforced with graphene fillers. Their findings demonstrate that the smooth material gradation in FGMs improves vibration performance and effectively eliminates the delamination problems typically observed in conventional composite materials. Priyadarshini et al. [21] utilized higher-order shear deformation theory to investigate the free vibration characteristics of epoxy-based functionally graded beams, incorporating axial material gradation modeled by a power law distribution. Using a Navier-type solution method, they accurately determined the natural frequencies and mode shapes, providing important insights for the dynamic analysis and optimal design of FGM structures.

In recent years, functionally graded material (FGM) beams have gained considerable attention in both research and engineering applications due to their superior mechanical performance, adaptability, and potential for optimizing structural behavior under complex loading conditions. To accurately investigate their responses, advanced finite element software such as ANSYS (19.5 R2, ANSYS, Inc., Canonsburg, Pennsylvania, USA) and ABAQUS has been extensively used, providing a flexible and reliable framework for simulating real-world scenarios that are often difficult to handle analytically. These tools allow researchers to study not only the static behavior, but also the dynamic characteristics, of FGM beams, including natural frequencies, mode shapes, and stress distributions, under various boundary conditions and geometric configurations. Unlike purely analytical approaches, which often require simplifying assumptions and can become cumbersome for complex geometries, combined loading, or geometrically nonlinear cases, finite element simulations enable the precise modeling of material gradation, layered structures, and irregular cross-sections.

Numerous studies in the literature have shown that numerical results obtained using ABAQUS or ANSYS closely match analytical solutions in simpler benchmark problems, while also allowing analyses of situations where manual calculations would be extremely time-consuming or practically infeasible. In addition, these simulations provide deeper insight into the effects of material gradation, shear deformation, and boundary conditions on the overall structural performance of FGM beams, which is valuable for both design and optimization purposes. ABAQUS and ANSYS are therefore widely adopted in the field, serving not only to validate analytical models but also to support the investigation of innovative design strategies for advanced composite structures. Their widespread use in the literature highlights their effectiveness in bridging the gap between theoretical predictions and practical engineering applications, underscoring their central role in contemporary research on functionally graded materials and smart structural systems. Hedia et al. [22] used ANSYS 2020 to perform a finite element analysis on functionally graded beams made of aluminum and alumina, analyzing the effects of different material distribution functions, namely, the power law, modified symmetric power law, sigmoid, and varying material index values on the resulting stress distribution. Abood and Al-Ansari [23] investigated the free vibration behavior of porous functionally graded beams (FGBs) under different support conditions using first-order shear deformation theory and ANSYS APDL 17.2. With the help of ANSYS Workbench, Özmen and Özhan [24] conducted analytical, computational, and experimental studies to investigate the bending and vibration behaviors of functionally graded beams modeled using the Euler–Bernoulli beam theory and fabricated via 3D printing. Khan et al. [25] carried out a free vibration analysis of functionally graded beams using the ABAQUS finite element software.

In this study, a functionally graded (FG) beam model (see Figure 1) was considered based on Timoshenko beam theory, also known as first-order shear deformation theory (FSDT), which extends classical Euler–Bernoulli beam theory by incorporating shear deformation effects that are neglected in the latter, allowing for a more accurate representation of short or thick beams under bending. The governing equations and boundary conditions were derived using Hamilton's principle, leading to Euler–Lagrange equations expressed in terms of three unknowns. For the analytical solution, we applied the Navier method, substituting trigonometric functions into the equations of motion to determine the natural frequencies for the first three modes (m = 1, 2, 3) of the FG Timoshenko beams. To make the results comparable with those in the existing literature, we presented the frequencies in a dimensionless form for simply supported boundary conditions in cases of $L/h = 5$ and $L/h = 20$. The Navier solution is widely recognized as a reliable analytical approach, especially for the static and dynamic analysis of functionally graded beams and plates. However, solving FG structures analytically can become very challenging when combined loading or geometric nonlinearities are involved, as the manual calculations become time-consuming or even impractical. In such cases, using advanced finite element software offers a much more convenient and efficient alternative.

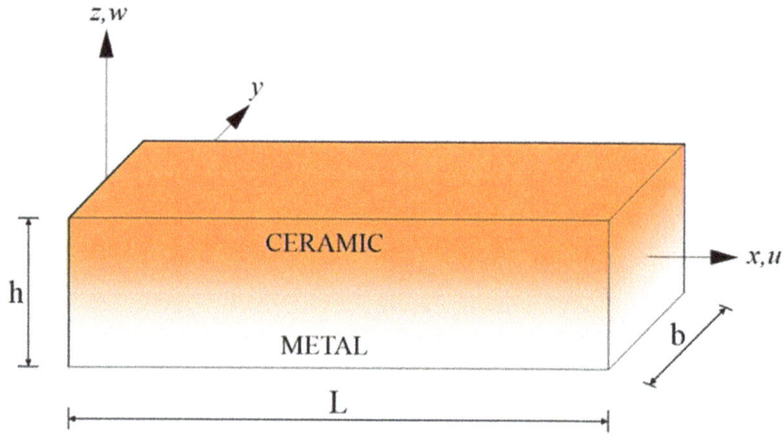

Figure 1. Model geometry of a functionally graded beam.

For the numerical analysis, ABAQUS was employed. To ensure accurate and convergent results, the beams were modeled using second-order hexahedral elements (HEX20/C3D20). Each layer's elastic modulus and density were defined according to the rule of mixtures, creating a smooth ceramic-to-metal material gradient along the beam. Natural frequencies were extracted using ABAQUS's Eigensolver with the Lanczos algorithm for different slenderness ratios. To compare with the analytical solution, the numerical frequencies were also converted into dimensionless form.

While many studies have explored the mechanical behavior of FG beams using different deformation theories, to the best of the authors' knowledge, no study has numerically modeled and solved FG beams in ABAQUS using both second-order hexahedral 3D solid elements (HEX20/C3D20) and 1D Timoshenko beam elements (B32). By examining both approaches on the same FG beam configurations, including variations in slenderness ratio and power–law index, we can directly compare the natural frequencies obtained from the 3D solid element model and the 1D Timoshenko beam model. Timoshenko B32 elements are particularly suitable for vibration analyses of short and thick beams because they account for both bending and shear deformations while keeping computational cost low. However, they offer lower accuracy compared to 3D solid elements like C3D20, and have limitations in modeling complex geometries and heterogeneous materials. This allows us to see which modeling approach converges most closely to the analytical solution. Such a comparison not only validates the numerical models, but also provides practical guidance on choosing the most suitable element type for capturing the dynamic behavior of FG beams. We believe that this study makes an important contribution to the literature, offering a clear reference for future investigations of FG beam vibrations and bridging the gap between detailed solid element modeling and efficient beam theory approaches.

2. Theory and Formulation

2.1. Material Properties

The variation in material properties is assumed to be governed by a power–law variation, defined as follows [5]:

$$P(z) = P_m V_m + P_c V_c \tag{1}$$

Here, P_m and P_c denote the material properties, while V_m and V_c represent volume fractions of the metal and ceramic associated with the given subindices, respectively.

$$V_m + V_c = 1 \tag{2}$$

The volume fraction of the ceramic phase, described by a power–law distribution, is expressed as follows:

$$V_c = \left(\frac{z}{h} + \frac{1}{2}\right)^p \tag{3}$$

Here, h is the beam height and p is a positive parameter known as the power–law index, ranging $0 \le p \le \infty$. When $p = 0$, the material is entirely ceramic. Using this relationship, the Young's modulus and mass density are determined by the following expressions:

$$\begin{aligned} E(z) &= E_m + (E_c - E_m)V_c \\ \rho(z) &= \rho_m + (\rho_c - \rho_m)V_c \end{aligned} \tag{4}$$

2.2. Constitutive Equations

The displacement field according to FSDT is expressed in a general form. It is important to note that, unlike the Euler–Bernoulli theory, the first-order shear component is also included in the field equations in this formulation [26],

$$u(x,z,t) = u_0(x,t) + z\xi(x,t) \tag{5}$$

$$w(x,z,t) = w_0(x,t) \tag{6}$$

where u and w show the axial and vertical displacement of the beam, and ξ demonstrates the angle of rotation of the normal to the mid-surface of the beam.

The non-zero linear strains derived from Equations (5) and (6) are the following:

$$\varepsilon_x = \frac{\partial u_0}{\partial x} + z\frac{\partial \xi}{\partial x} \tag{7}$$

$$\gamma_{xz} = \frac{\partial w_0}{\partial x} + \xi \tag{8}$$

Assuming the material of the functionally graded (FG) beam obeys Hooke's law, the stress components within the Timoshenko beam are expressed in matrix form as

$$\begin{Bmatrix} \sigma_x \\ \tau_{xz} \end{Bmatrix} = \begin{bmatrix} Q_{11} & 0 \\ 0 & Q_{55} \end{bmatrix} \begin{Bmatrix} \varepsilon_x \\ \gamma_{xz} \end{Bmatrix} \tag{9}$$

where

$$Q_{11} = E(z),\ Q_{55} = \frac{E(z)}{2(1+\nu)} \tag{10}$$

2.3. Equations of Motion

The equations of motion are derived using Hamilton's principle by integrating over time, as follows [27]:

$$\delta \int_{t_1}^{t_2} (U + V - T)dt = 0 \tag{11}$$

Here, δU, δV, and δT represent the virtual variations of the strain energy, potential energy, and kinetic energy, respectively.

The variational form of the strain energy, δU, is given by:

$$\delta U = \int_0^L \int_{-\frac{h}{2}}^{\frac{h}{2}} (\sigma_x \delta \varepsilon_x + \tau_{xz} \delta \gamma_{xz}) dz dx$$
$$\delta U = \int_0^L \left[N_x \frac{\partial \delta u_0}{\partial x} + M_x \frac{\partial \delta \xi}{\partial x} + Q_{xz} \left(\frac{\partial \delta w_0}{\partial x} + \delta \xi \right) \right] dx \quad (12)$$

Accordingly, the stress resultants N_x, M_x, and Q_{xz} are defined as

$$\{N_x, M_x, Q_{xz}\} = \int_{-\frac{h}{2}}^{\frac{h}{2}} \sigma_x (1, z, k_s \tau_{xz}) dz \quad (13)$$

where k_s indicates shear coefficient.

The work done by the external vertical load q can be expressed as follows:

$$\delta V = \int q \delta w dA \quad (14)$$

The kinetic energy δT can be written as follows [28]:

$$T = \frac{1}{2} \int_0^L \int_A \left(\rho(z) \left(\dot{u}^2 + \dot{w}^2 \right) \right) dA dx$$
$$\delta T = \int_0^L \int_{-\frac{h}{2}}^{\frac{h}{2}} (\rho(z) (\dot{u} \delta u + \dot{w} \delta w)) dz dx \quad (15)$$

By substituting the variational expressions of δU, δV, and δT, presented in Equations (12), (14) and (15), respectively, into Equation (11), the governing equations of motion for the functionally graded beam are obtained as follows:

$$\begin{aligned} \delta u_0 &: N'_x = I_0 \ddot{u}_0 + I_1 \ddot{\xi} \\ \delta \xi &: M'_x - Q_{xz} = I_1 \ddot{u}_0 + I_2 \ddot{\xi} \\ \delta w_0 &: Q'_{xz} + q = I_0 \ddot{w}_0 \end{aligned} \quad (16)$$

Here, the double dots above the parameters denote the components derived from the kinetic energy, representing the second derivatives with respect to time. The inertial coefficients are defined as follows:

$$\{I_0, I_1, I_2\} = \int_A \rho(z) \left(1, z, z^2 \right) dA \quad (17)$$

By reformulating Equation (16) in conjunction with Equation (17), the governing equations of motion for the functionally graded material beam are comprehensively articulated and expressed as

$$A_{11} \frac{\partial^2 u_0}{\partial x^2} + B_{11} \frac{\partial^2 \xi}{\partial x^2} - I_0 \frac{\partial^2 u_0}{\partial t^2} - I_1 \frac{\partial^2 \xi}{\partial t^2} = 0 \quad (18)$$

$$B_{11} \frac{\partial^2 u_0}{\partial x^2} + D_{11} \frac{\partial^2 \xi}{\partial x^2} - A_{55} \left[\frac{\partial w_0}{\partial x} + \xi \right] - I_1 \frac{\partial^2 u_0}{\partial t^2} - I_2 \frac{\partial^2 \xi}{\partial t^2} = 0 \quad (19)$$

$$A_{55}\left[\frac{\partial^2 w_0}{\partial x^2}+\frac{\partial \xi}{\partial x}\right]+q-I_0\frac{\partial^2 w_0}{\partial t^2}=0 \tag{20}$$

where

$$\{A_{11},B_{11},D_{11}\}=\int_{-\frac{h}{2}}^{\frac{h}{2}} Q_{11}\left(1,z,z^2\right)dz \tag{21}$$

$$A_{55}=\int_{-\frac{h}{2}}^{\frac{h}{2}} k_s Q_{55} dz \tag{22}$$

3. Solution Procedure

3.1. Analytical Part of Solution

In this section, an analytical solution to the governing equations for the free vibration of a supported functionally graded beam is presented, based on the Navier method. The transverse load distribution is assumed to be zero. To this end, the displacement functions are expressed as the product of undetermined coefficients and known trigonometric functions, ensuring that the governing equations and boundary conditions are satisfied at $x = 0$ and $x = L$ [28],

$$\left\{\begin{array}{c} u_0 \\ \xi \\ w_0 \end{array}\right\}=\sum_{m=1}^{\infty}\left\{\begin{array}{c} U_m e^{i\omega t}Cos(\lambda x) \\ \xi_m e^{i\omega t}Cos(\lambda x) \\ W_m e^{i\omega t}\text{Sin}(\lambda x) \end{array}\right\} \tag{23}$$

where the unknown Fourier coefficients U_m, ξ_m, and W_m should be determined; w is the eigenfrequency associated with mth eigenmode, and $\lambda=\frac{m\pi}{L}$.

For the simply supported beam, essential and natural boundary conditions are given as shown below [29]:

$$\begin{array}{l} N_x\delta u_0|_0^L=0 \\ M_x\delta\xi|_0^L=0 \\ Q_{xz}\delta w_0|_0^L=0 \\ u=w=0 \quad at \quad x=0,L \end{array} \tag{24}$$

The following q load is considered as in study [27]:

$$q(x)=\sum_{m=1}^{\infty} q_m \sin(\alpha x) \tag{25}$$

$$q_m=\frac{2}{L}\int_0^L q(x)\sin(\alpha x)dx \tag{26}$$

$$q_m=\left\{\begin{array}{ll} q_0(m=1) & \text{for sin usoidalload } q_0 \\ \frac{4q_0}{m\pi}(m=1,3,5,\ldots) & \text{for uniform load } q_0 \end{array}\right. \tag{27}$$

By substituting the variables u_0, ξ, and w_0 from Equation (23) into the equations of motion, analytical solutions are obtained by evaluating the determinant of the matrix formed by the stiffness components K_{ij} and mass components m_{ij} [28]. In the present study, q_m is assumed to be zero (q_m = 0).

$$\left(\begin{bmatrix} K_{11} & K_{12} & K_{13} \\ K_{12} & K_{22} & K_{23} \\ K_{13} & K_{23} & K_{33} \end{bmatrix} - \omega^2 \begin{bmatrix} m_{11} & m_{12} & m_{13} \\ m_{12} & m_{22} & m_{23} \\ m_{13} & m_{23} & m_{33} \end{bmatrix}\right) \begin{Bmatrix} U_m \\ \xi_m \\ W_m \end{Bmatrix} = \begin{Bmatrix} 0 \\ 0 \\ q_m \end{Bmatrix} \tag{28}$$

where

$$\begin{aligned} &K_{11} = -A_{11}\lambda^2, \quad K_{12} = K_{21} = -B_{11}\lambda^2, \quad K_{13} = K_{31} = 0, \\ &K_{22} = -A_{55} - D_{11}\lambda^2, \quad K_{23} = K_{32} = -A_{55}\lambda, \quad K_{33} = -A_{55}\lambda^2, \\ &m_{11} = -I_0, \quad m_{12} = -I_1, \quad m_{13} = 0, \\ &m_{22} = -I_2, \quad m_{23} = 0, \quad m_{33} = -I_0 \end{aligned} \tag{29}$$

3.2. Numerical Part

The commercial finite element package ABAQUS was employed to perform the numerical simulations. Pre-processing and post-processing tasks were conducted using ABAQUS/CAE (2023, Dassault Systèmes, Providence, RI, USA) [30], whereas the actual solution phase was carried out with the ABAQUS solver. To compute the natural frequencies of the models, a frequency extraction analysis was performed using the Lanczos eigensolver. The Lanczos method is widely recognized as a reliable approach for determining natural frequencies and is routinely applied in both modal analyses and frequency response function (FRF) evaluations. Finite element analysis (FEA) provides the ability to investigate structural responses for complex geometries, loading conditions, and material distributions. However, it is important to emphasize that FEA always produces approximate solutions. The accuracy of these solutions depends heavily on both mesh refinement and element type selection. In general, mesh convergence is improved by reducing element size and increasing the number of elements. Yet, excessively fine meshes lead to a significant increase in computational cost. Therefore, an optimal balance between accuracy and efficiency must be achieved. Moreover, the correct choice of element formulation is essential. In certain cases, first-order elements provide adequate accuracy, while in others, second-order elements are more appropriate. For solid models, hexahedral elements are usually preferred due to their numerical stability and convergence properties, while tetrahedral elements are adopted when handling irregular or complicated geometries. Similarly, the correct decision between solid, shell, and beam elements must be made depending on the geometry and structural characteristics under investigation. In the present study, two separate modeling strategies were adopted in ABAQUS to evaluate the natural frequencies of functionally graded beams. The first approach involved modeling the beams using 3D solid elements, while the second approach relied on 1D beam elements. For the solid element-based model, a second-order hexahedral mesh (HEX20/C3D20) was chosen to ensure analytical convergence and accuracy. Each element dimension was set to 5 × 5 × 5 mm, which provided a sufficient level of mesh quality. For the slender beam with a slenderness ratio of L/h = 20, the dimensions were 0.1 × 0.1 × 2 m, whereas for the short beam with L/h = 5, the dimensions were 0.1 × 0.1 × 0.5 m. To account for the material variation along the cross-section of the beam, the model was divided into 10 groups of elements. Each group was assigned its own material property set, allowing the smooth transition from metal to ceramic across the beam thickness to be accurately captured. Different power–law index (p) values were incorporated by defining separate material models for each case. To represent the simply supported boundary condition, rigid body elements were introduced at both ends of the beam. These rigid bodies were assigned appropriate constraints to their reference nodes, enabling one end to be fixed while allowing sliding at the other. Such an approach ensured the free rotation of the beam ends, consistent with the simply supported condition. The final solid-element-based finite element models contained approximately 160,000 hexahedral elements, two rigid body elements, and a total of 690,083 nodes for $L/h = 20$, and approximately 40,000 hexahedral elements, two rigid body elements, and 173,482 nodes for $L/h = 5$. In both cases, the ABAQUS solver was instructed to extract the first 10 natural frequencies.

A second modeling approach was carried out using 1D beam elements. Beam elements are generally advantageous when dealing with simple geometries, as they significantly reduce the number of degrees of freedom and computational effort. However, in the context of functionally graded materials (FGMs), additional care is required because the mechanical properties vary continuously within the material. This allows material transitions to be modeled accurately within individual elements, even in a 1D beam formulation. To achieve reliable and convergent results, the B32 Timoshenko beam element was selected. Timoshenko beam theory accounts for both bending and shear deformation, making it suitable for beams with lower slenderness ratios as well as for FGMs where shear flexibility cannot be neglected. A mesh size of 5 mm was employed to ensure the adequate quality and smooth distribution of material properties along the beam length. For $L/h = 20$, the dimensions were $0.1 \times 0.1 \times 2$ m, discretized into 400 B32 elements, two rigid body elements, and 801 nodes, while for $L/h = 5$, the dimensions were $0.1 \times 0.1 \times 0.5$ m, discretized into "100" B32 elements, two rigid body elements, and 201 nodes. As in the solid element case, the solver was requested to compute the first 10 natural frequencies. In summary, the study employed two complementary finite element approaches, a detailed 3D solid element model and a computationally efficient 1D beam element model to investigate the natural frequencies of functionally graded beams. While the 3D solid model captures the geometry and material distribution in greater detail, the beam element approach provides significant reductions in solution time while maintaining accuracy through the use of Timoshenko beam theory. The combination of these methods allows for both accuracy verification and computational efficiency, ensuring that the analysis of FGMs is both reliable and practical.

The modeling steps of the FGM beam in ABAQUS are presented as seen in Figure 2.

(**a**) First, the material properties were manually calculated for each p value. Material characteristics, including Young's modulus, density, and Poisson's ratio, were defined using the Elastic Material keyword. An appropriate mesh size was determined through a convergence study. After meshing, the material properties were assigned accordingly.
In the 3D model, ten distinct material definitions were applied to separate mesh regions. In contrast, the 1D model utilized a single material definition with internal transitions between materials. Two separate models were developed for aspect ratios $L/h = 5$ and $L/h = 20$.

↓

(**b**) Rigid body elements were defined on both ends of the beam to represent the boundary conditions accurately. For simply supported beam configurations, rigid body modeling is essential to ensure realistic and accurate simulation results.

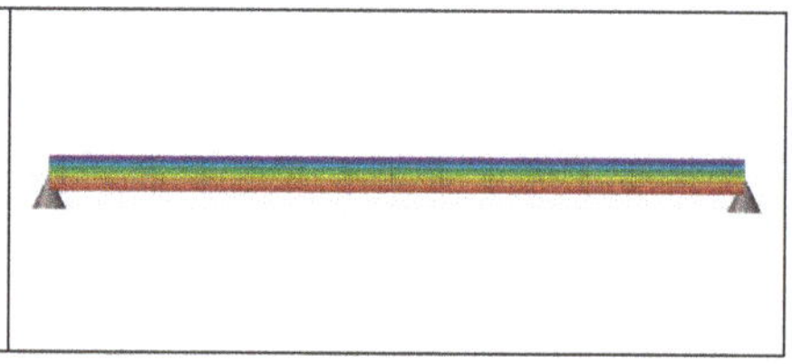

↓

(**c**) After defining the boundary conditions in the third step, the mesh of the FGM beam was assigned.

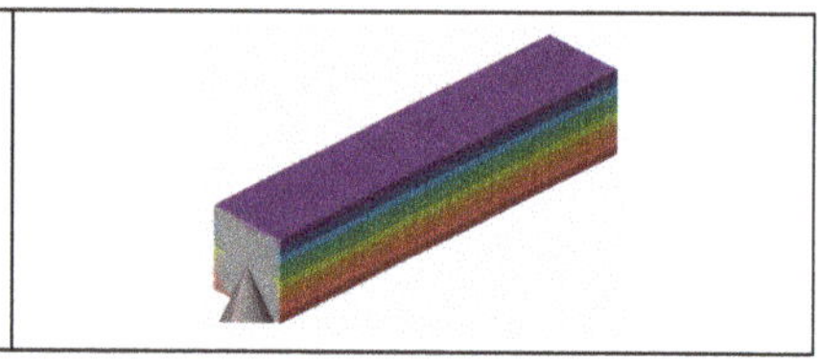

↓

(**d**) Modal analysis was carried out using the ABAQUS finite element solver. For each value of p, the corresponding natural frequencies and mode shapes were extracted.

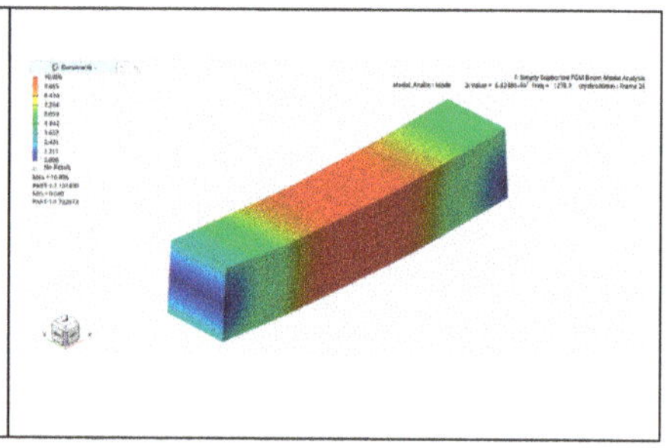

Figure 2. Flowchart for the modeling of the FGM beam.

4. Results and Discussion

The free vibration behavior of the first mode of FG Timoshenko beams was analyzed for two different length-to-height ratios, using the material properties listed in Table 1 as a reference.

Table 1. Material properties of the FGM beam [27,31].

Material	*Young Modulus, E (GPa)*	*Density, ρ (kg/m³)*	*Poisson Ratio, υ*
Alumina (Al_2O_3)	380	3960	0.3
Aluminum (Al)	70	2702	0.3

For different power–law index values, the results obtained from analytical and numerical solutions are presented in Tables 2–4, plotted as seen Figures 3 and 4 and compared with those reported in the literature.

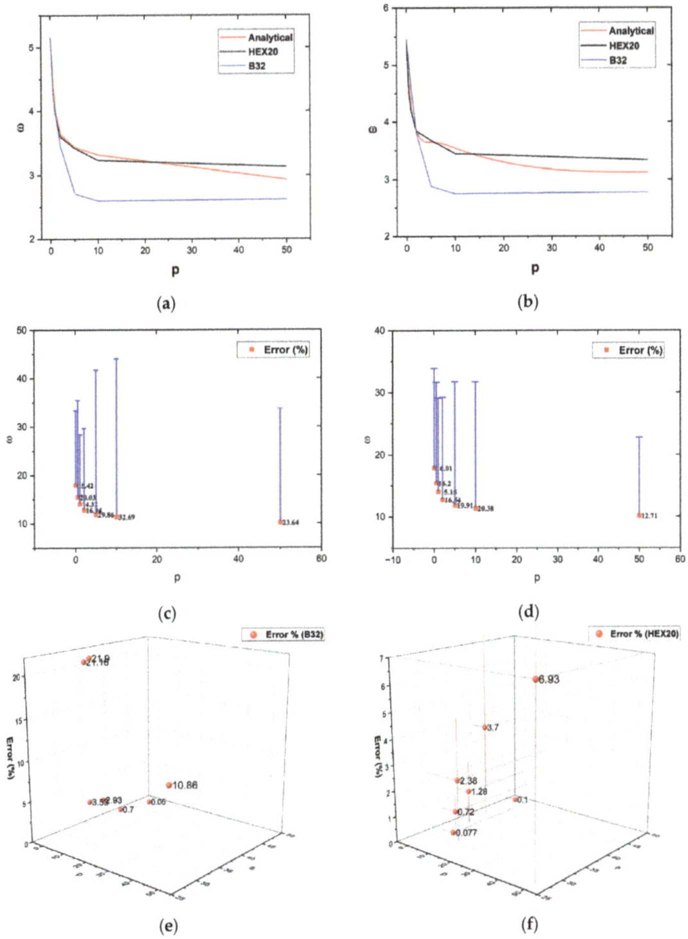

Figure 3. (**a**) Comparison of dimensionless frequencies values obtained from analytical and numerical solutions for the $L/h = 5$ (m = 1). (**b**) Comparison of dimensionless frequencies values obtained from analytical and numerical solutions for the $L/h = 20$ (m = 1). (**c**) Percentage error of the dimensionless frequency obtained from the numerical results (B32) compared to the analytical reference results ($L/h = 5$, m = 2). (**d**) Percentage error of the dimensionless frequency obtained from the numerical results (HEX20) compared to the analytical reference results ($L/h = 5$, m = 2). (**e**) Three-dimensional percentage error of the dimensionless frequency obtained from the numerical results (B32) compared to the analytical reference results ($L/h = 20$, m = 3). (**f**) Three-dimensional percentage error of the dimensionless frequency obtained from the numerical results (HEX20) compared to the analytical reference results ($L/h = 20$, m = 3).

Figure 4. ABAQUS model of FG beam (HEX20).

The dimensionless frequency can be calculated using the following formula [32]:

$$\overline{\omega} = \left(\frac{\omega L^2}{h}\right)\sqrt{\frac{\rho_m}{E_m}} \tag{30}$$

Table 2. Comparison of dimensionless natural frequencies of simply supported FG beams for the first mode.

L/h	*p*	*Present (TBT)*	*ABAQUS (HEX20)*	*ABAQUS (Timoshenko 1D)*	*Chen* et al. [33]	*CBT* [31]
5	0	5.1524	5.1563	5.1569	5.1525	5.3953
	0.5	4.4078	4.2450	4.2354	4.4079	4.5931
	1	3.9902	3.9370	3.9623	3.9904	4.1484
	2	3.6343	3.6021	3.4407	3.6346	3.7793
	5	3.4312	3.4200	2.7061	3.4315	3.5949
	10	3.3134	3.2315	2.5960	3.3136	3.4921
	50	2.9288	3.1269	2.6182	-	-
20	0	5.4603	5.4612	5.4607	5.4603	5.4777
	0.5	4.6509	4.4791	5.0114	4.6509	4.6641
	1	4.2050	4.1685	4.5509	4.2051	4.2163
	2	3.8367	3.8463	3.8015	3.8368	3.8472
	5	3.6508	3.6858	2.8814	3.6509	3.6628
	10	3.5415	3.4493	2.7494	3.5416	3.5547
	50	3.1207	3.3338	2.7724	-	-

Table 3. Comparison of dimensionless natural frequencies of simply supported FG beams for the second mode.

L/h	*p*	*Present (TBT)*	*ABAQUS (HEX20)*	*ABAQUS (Timoshenko 1D)*	*Chen* et al. [33]	*CBT* [27]
5	0	17.8711	15.0112	15.1155	17.8711	20.6187
	0.5	15.4263	12.9269	12.3368	15.4277	17.5415
	1	14.0030	11.8785	11.9982	14.0064	15.7982
	2	12.7120	10.6079	10.5592	12.7179	14.3260
	5	11.8157	9.4636	8.2881	11.8226	13.5876
	10	11.3073	9.0001	7.6107	11.3119	13.2376
	50	10.0500	8.7717	7.6743	-	-

Table 3. *Cont.*

L/h	p	*Present (TBT)*	*ABAQUS (HEX20)*	*ABAQUS (Timoshenko 1D)*	*Chen* et al. [33]	*CBT* [27]
20	0	21.5732	21.5831	21.5782	21.5732	21.8438
	0.5	18.3931	17.7203	19.2787	18.3931	18.5987
	1	16.6344	16.4784	17.6408	16.6344	16.8100
	2	15.1715	15.1744	14.8756	15.1715	15.3334
	5	14.4109	14.5201	11.3722	14.4110	14.5959
	10	13.9652	13.6022	10.8641	13.9653	14.1676
	50	12.3146	13.1533	10.9555	-	-

Table 4. Comparison of dimensionless natural frequencies of simply supported FG beams for the third mode.

L/h	p	*Present (TBT)*	*ABAQUS (HEX20)*	*ABAQUS (Timoshenko 1D)*	*Chen* et al. [33]	*CBT* [27]
5	0	34.1449	34.2776	34.2807	34.1449	43.3483
	0.5	29.7166	28.9559	20.8563	29.7231	36.8308
	1	27.0525	26.8252	20.7280	27.0683	33.0278
	2	24.4970	23.9218	19.8466	24.5242	29.7458
	5	22.4642	20.9403	17.6138	22.4933	28.0850
	10	21.3219	20.9403	17.2456	21.3399	27.4752
	50	19.0238	19.3222	17.4048	-	-
20	0	47.5921	47.6399	47.6167	47.5921	48.8999
	0.5	40.6377	39.1212	40.9196	40.6378	41.6328
	1	36.7673	36.2973	37.8398	36.7677	37.6173
	2	33.5135	33.2699	32.3465	33.5142	34.2954
	5	31.7473	31.7229	25.0470	31.7482	32.6357
	10	30.7174	29.9764	23.9720	30.7180	31.6883
	50	27.1162	28.9962	24.1755	-	-

5. Conclusions

In this study, the free vibration behavior of functionally graded (FG) beams was investigated using both analytical formulations and numerical simulations (ABAQUS solver). The analysis was carried out within the framework of Timoshenko beam theory, also known as the first-order shear deformation theory (FSDT). The primary objective was to demonstrate that, although FG structures are often challenging to analyze analytically, finite element software can provide an effective and reliable tool for addressing such problems. To this end, the governing equations were derived by incorporating shear effects and applying the principle of minimum potential energy, and the resulting dimensionless natural frequencies were obtained and compared with those from numerical models.

Two different modeling strategies were carried out in ABAQUS alongside the analytical approach. A HEX20 solid element model was built where the material gradation was defined layer by layer, while a B32 Timoshenko beam element was used to develop a beam-type 1D model. The free vibration frequencies were calculated for the first three modes under simply supported boundary conditions, considering two slenderness ratios (L/h = 5 and L/h = 20) and a range of power–law index (p) values. Results were also obtained for p = 50, which have not been reported in the literature so far. To the best of our knowledge, no other study has modeled both solid and 1D elements in ABAQUS and compared them systematically with each other, as well as with the analytical solution, which highlights the originality of this work.

As can be seen from Tables 2–4 and Figure 3, the results for all three modes indicate that the HEX20 solid element model shows consistent agreement with the analytical solutions. In contrast, the frequencies obtained from the B32 Timoshenko beam element exhibit slightly larger deviations from the analytical results compared to the HEX20 model, especially beyond a certain p value. This behavior is thought to stem from how the 1D element represents gradation transition, which introduces irregularities in the results. In addition, Figures 5–10 present the frequencies obtained for both the HEX20 solid element and the B32 beam element, corresponding to the first three modes at a selected value of p. Overall, the authors believe that the outcomes of this study can serve as a practical reference for the computational analysis of FGM structures, and offer guidance for future research in the field.

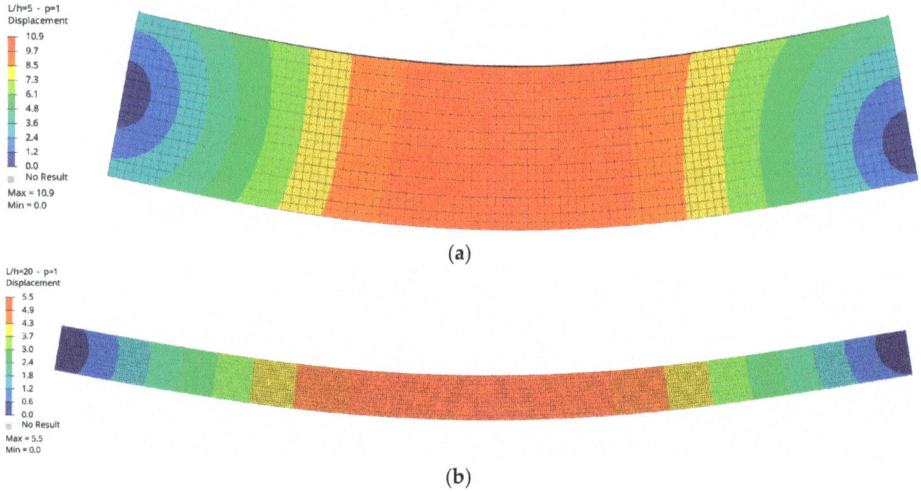

Figure 5. The first mode of the FG beam (**a**) for the length-to-height ratio (L/h = 5) and p = 1, and (**b**) for (L/h = 20) and p = 1 (HEX20).

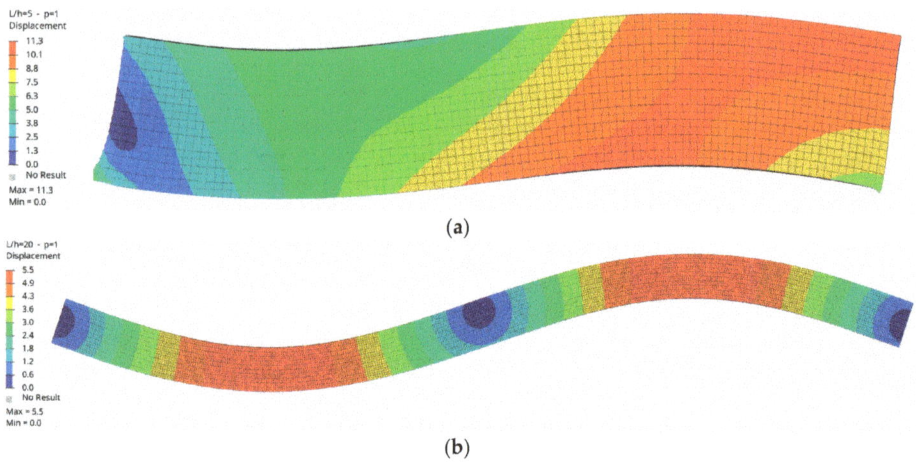

Figure 6. The second mode of the FG beam (**a**) for the length-to-height ratio (L/h = 5) and p = 1, and (**b**) for (L/h = 20) and p = 1 (HEX20).

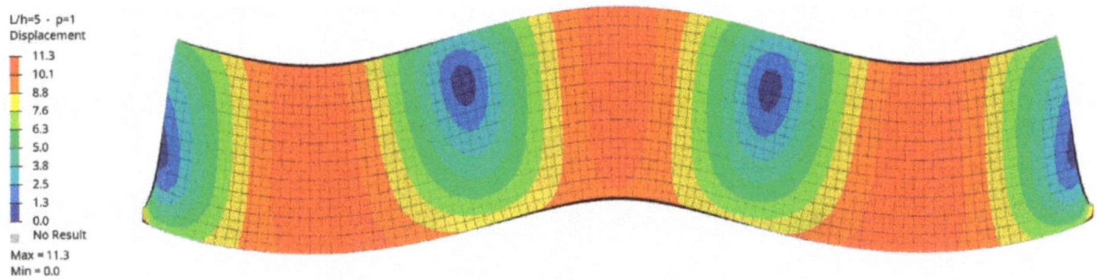

(**a**)

(**b**)

Figure 7. The third mode of the FG beam (**a**) for the length-to-height ratio (L/h = 5) and p = 1, and (**b**) for (L/h = 20) and p = 1 (HEX20).

(**a**)

(**b**)

Figure 8. The first mode of the FG beam (**a**) for the length-to-height ratio (L/h = 5) and p = 1, and (**b**) for (L/h = 20) and p = 1 (Timoshenko beam).

(**a**)

Figure 9. *Cont.*

(**b**)

Figure 9. The second mode of the FG beam (**a**) for the length-to-height ratio ($L/h = 5$) and $p = 1$, and (**b**) for ($L/h = 20$) and $p = 1$ (Timoshenko beam).

(**a**)

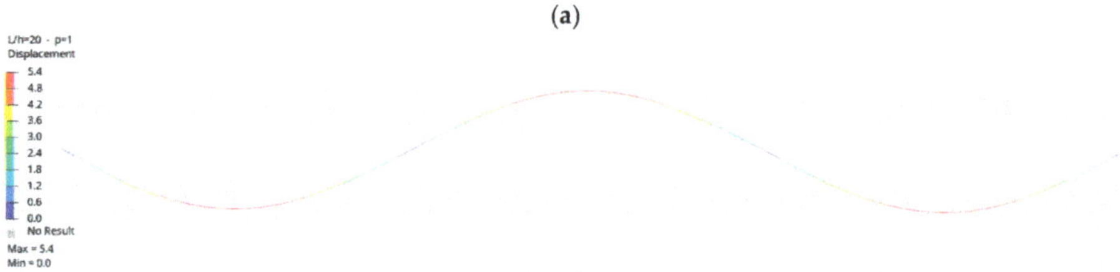

(**b**)

Figure 10. The third mode of the FG beam (**a**) for the length-to-height ratio ($L/h = 5$) and $p = 1$, and (**b**) for ($L/h = 20$) and $p = 1$ (Timoshenko beam).

Author Contributions: Methodology, M.Ç.; Software, A.F.Y.; Validation, A.F.Y.; Formal analysis, M.Ç.; Investigation, M.Ç. and E.D.; Resources, A.F.Y.; Writing—original draft, E.D.; Writing—review and editing, M.Ç.; Supervision, M.Ç. All authors have read and agreed to the published version of the manuscript.

Funding: This research received no external funding.

Institutional Review Board Statement: Not applicable.

Informed Consent Statement: Not applicable.

Data Availability Statement: The data presented in this study are available on request from the corresponding author.

Conflicts of Interest: The authors declare no conflicts of interest.

References

1. Qian, L.F.; Batra, R.C.; Chen, L.M. Static and dynamic deformations of thick functionally graded elastic plates by using higher-order shear and normal deformable plate theory and meshless local Petrov–Galerkin method. *Compos. Part B Eng.* **2004**, *35*, 685–697. [CrossRef]
2. Reddy, J. *Thermomechanical Behavior of Functionally Graded Materials*, 1st ed.; Department of Mechanical Engineering, Texas A&M University: College Station, TX, USA, 1998; pp. 3–20.
3. Rahmani, A.; Babaei, A.; Faroughi, S. Vibration Characteristics of Functionally Graded Micro-Beam Carrying an Attached Mass. *Mech. Pf Adv. Compos. Struct.* **2020**, *7*, 49–58.

4. Sedighi, H.M.; Keivani, M.; Abadyan, M. Modified continuum model for stability analysis of asymmetric FGM double-sided NEMS: Corrections due to finite conductivity, surface energy and nonlocal effect. *Compos. Part B Eng.* **2015**, *83*, 117–133. [CrossRef]
5. Alshorbagy, A.E.; Eltaher, M.A.; Mahmoud, F.F. Free vibration characteristics of a functionally graded beam by finite element method. *Appl. Math. Model.* **2011**, *35*, 412–425. [CrossRef]
6. Avcar, M. Free vibration of imperfect sigmoid and power law functionally graded beams. *Steel Compos. Struct.* **2019**, *30*, 603–615.
7. Kahya, V.; Turan, M. Finite element model for vibration and buckling of functionally graded beams based on the first-order shear deformation theory. *Compos. Part B Eng.* **2017**, *109*, 108–115. [CrossRef]
8. Çelik, M.; Artan, R. An investigation of static bending of a bi-directional strain-gradient Euler–Bernoulli nano-beams with the method of initial values. *Microsyst. Technol.* **2020**, *26*, 2921–2929. [CrossRef]
9. Kareem, M.G.; Sadiq, S.E.; Al-Raheem, S.K.; Alansari, L.S. Analysis the free vibration of functionally graded material plate by using new displacement function. *Results Eng.* **2025**, *25*, 103756. [CrossRef]
10. Su, X.S.; Hu, T.; Zhang, W.; Kang, H.; Cong, Y.; Yuan, Q. Transfer matrix method for free and forced vibrations of multi-level functionally graded material stepped beams with different boundary conditions. *Appl. Math. Mech.* **2024**, *45*, 983–1000. [CrossRef]
11. El Hantati, I.; Ouyassafte, O.; El Khouddar, Y.; Belhaou, M.; Adri, A.; Benamar, R. Analysis of the transverse vibration of a multistepped FGM beam resting on a Winkler foundation in a thermal environment and carrying concentrated masses. *Results Eng.* **2024**, *23*, 102822. [CrossRef]
12. Wang, Y.; Liu, B.; Zhu, J.; Zhou, W.; Jiang, L.; Pan, C.; Xu, J. Thermal vibration of stiffened FGM plates with cutouts using Nitsche-based isogeometric approach. *Thin-Walled Struct.* **2025**, *210*, 113026. [CrossRef]
13. Hilali, Y.; Rassam, M.; Mesmoudi, S.; Sitli, Y.; Elmhaia, O.; Rammane, M.; Askour, O.; Bourihane, O. A high-order approach for thermal buckling and post-buckling analysis of functionally graded sandwich beams. *Acta Mech.* **2025**, *236*, 3543–3563. [CrossRef]
14. Pradhan, K.K.; Chakraverty, S. Free vibration of Euler and Timoshenko functionally graded beams by Rayleigh–Ritz method. *Compos. Part B Eng.* **2013**, *51*, 175–184. [CrossRef]
15. Yıldırım, B. Investigation of thermal shock fracture in an edge-cracked functionally graded layer using finite element method. *J. Fac. Eng. Archit. Gazi Univ.* **2004**, *19*, 235–245.
16. Kahya, V.; Turan, M. Vibration and stability analysis of functionally graded sandwich beams by a multi-layer finite element. *Compos. Part B Eng.* **2018**, *146*, 198–212. [CrossRef]
17. Nguyen, T.K.; Nguyen, B.D. A new higher-order shear deformation theory for static, buckling and free vibration analysis of functionally graded sandwich beams. *J. Sandw. Struct. Mater.* **2015**, *17*, 613–631. [CrossRef]
18. Echouai, B.; Adri, A.; El Khouddar, Y.; Outassafte, O.; El Hantati, I.; Echouai, E.K.; Hassa, M.; Benamar, R. Analysis of transverse vibrations of functionally graded beams with magnetostrictive Terfenol-D layers resting on an elastic foundation. *Sci. Afr.* **2025**, *28*, e02651. [CrossRef]
19. Benadouda, M.; Bourouis, M.E.A.; Dahmane, M.; Bennai, R.; Atmane, H.A.; Safer, O. Dynamic response of wave propagation in functionally graded beams with defects: Effects of porosity and cracks. *Acta Mech.* **2025**, *236*, 2279–2296. [CrossRef]
20. Jayachandiran, G.; Ramamoorthy, M. Free and forced vibration analysis of functionally graded composite beam with graphene filler reinforcement—Experimental and simulation study. *Polym. Compos.* **2025**, *46*, S70–S85. [CrossRef]
21. Priyadarshini, A.; Sutar, M.K.; Pattnaik, S. Refined Higher-Order Shear Deformation Analysis of Axial Functionally Graded Beams With Nano-Graphene Reinforcement. *J. Vib. Eng. Technol.* **2025**, *13*, 236. [CrossRef]
22. Hedia, H.S.; Almas, M.A.; Attar, H.M.; Hedia, M.H.; Gadelmawla, E.S.; Soliman, M.A. Numerical and analytical stress analysis for a FGM beam. *Mater. Test.* **2025**, *67*, 1046–1055. [CrossRef]
23. Abood, N.K.; Al-Ansari, L.S. Impact of porosity distribution on vibration of porous functionally graded beams. *Math. Model. Eng. Probl.* **2025**, *12*, 687–708. [CrossRef]
24. Özmen, U.; Özhan, B.B. Mechanics of functionally graded beams: Analytical, computational, and experimental analyses. *Int. J. Mech. Mater. Des.* **2025**, *21*, 577–590. [CrossRef]
25. Khan, A.A.; Alam, M.N.; ur Rahman, N.; Wajid, M. Finite element modelling for static and free vibration response of functionally graded beam. *Lat. Am. J. Solids Struct.* **2016**, *13*, 690–714. [CrossRef]
26. Chakraborty, A.; Gopalakrishnan, S.; Reddy, J.N. A new beam finite element for the analysis of functionally graded materials. *Int. J. Mech. Sci.* **2003**, *45*, 519–539. [CrossRef]
27. Thai, H.T.; Vo, T.P. Bending and free vibration of functionally graded beams using various higher-order shear deformation beam theories. *Int. J. Mech. Sci.* **2012**, *62*, 57–66. [CrossRef]
28. Larbi, O.L.; Kaci, A.; Houari, M.S.A.; Tounsi, A. An Efficient Shear Deformation Beam Theory Based on Neutral Surface Position for Bending and Free Vibration of Functionally Graded Beams. *Mech. Based Des. Struct. Mach.* **2013**, *41*, 421–433. [CrossRef]
29. Aydogdu, M.; Taskin, V. Free vibration analysis of functionally graded beams with simply supported edges. *Mater. Des.* **2007**, *28*, 1651–1656. [CrossRef]
30. Dassault Systemes, "ABAQUS FEA". 2024. Available online: https://www.3ds.com/products/simulia/abaqus (accessed on 20 August 2025).

31. Çelik, M.; Gündoğdu, E.; Özdilek, E.E.; Demirkan, E.; Artan, R. Artificial Neural Net-work (ANN) Validation Research: Free Vibration Analysis of Functionally Graded Beam via Higher-Order Shear Deformation Theory and Artificial Neural Network Method. *Appl. Sci.* **2024**, *14*, 217. [CrossRef]
32. Zohra, Z.; Lemya, H.; Abderahman, Y.; Mustapha, M.; Abdelouahed, T.; Djamel, O. Free vibration analysis of functionally graded beams using a higher-order shear deformation theory. *Math. Model. Eng. Probl.* **2017**, *4*, 7–12. [CrossRef]
33. Chen, W.R.; Chang, H. Vibration Analysis of Functionally Graded Timoshenko Beams. *Int. J. Struct. Stab. Dyn.* **2018**, *18*, 185007. [CrossRef]

Disclaimer/Publisher's Note: The statements, opinions and data contained in all publications are solely those of the individual author(s) and contributor(s) and not of MDPI and/or the editor(s). MDPI and/or the editor(s) disclaim responsibility for any injury to people or property resulting from any ideas, methods, instructions or products referred to in the content.

Journal of
Composites Science

MDPI

Article

Effect of Rubber Fiber Content on the Mechanical Properties of Calcareous Sand

Yuzhu Cheng [1], Hansheng Geng [2,*], Lei Wang [3,*], Yang Wang [4], Guoyue Yang [4], Yongsheng Xie [4], Linjian Ma [2] and Chun Li [5]

1 College of Civil Engineering, Hunan University of Science and Technology, Xiangtan 411201, China
2 State Key Laboratory of Disaster Prevention and Mitigation of Explosion and Impact, Army Engineering University of PLA, Nanjing 210000, China
3 School of Civil Engineering and Architecture Engineering, Guilin University of Technology, Guilin 541000, China
4 College of Civil Engineering and Mechanics, Xiangtan University, Xiangtan 411100, China
5 Natural Resources Bureau of Pingjiang County, Pingjiang 414500, China
* Correspondence: hsgeng@aeu.edu.cn (H.G.); wanglei@glut.edu.cn (L.W.)

Abstract: The application of rubber in geotechnical engineering has gained widespread popularity due to its potential to enhance the engineering properties of foundation fills while reducing environmental pollution. This study focuses on investigating the influence of the rubber fiber content on the performance of calcareous sand by conducting a series of triaxial tests. The effects of the rubber fiber content and axial pressure on the strength, deformation, permeability, and particle breakage of rubber–calcareous sand were systematically studied. The experimental results reveal that increasing the rubber fiber content reduces the strength of rubber–calcareous sand, but it also inhibits the shear dilation and mitigates the occurrence of rupture surfaces: the sample with a rubber content of more than 10% only has shear-contraction. Both the rubber fiber content and axial stress contribute to the increased impermeability of rubber-modified calcareous sand, although they exhibit different characteristics. The relationship between the rubber fiber content and permeability coefficient is linear, while, under increasing axial stress, the permeability coefficient initially decreases rapidly; when the deviatoric stresses exceeds 1000 kPa, the decreasing rate slows down. Furthermore, rubber fiber significantly reduces particle breakage in calcareous sand. The relationship between the input energy applied to rubber-modified calcareous sand and the relative breakage rate of calcareous sand can be well-fitted with a power function. Samples with a higher rubber fiber content exhibit a lower relative breakage rate of calcareous sand under the same absorbed input energy. Through the research results of this paper, the best rubber ratio can be selected as the road filler in engineering practice to ensure both cost-effectiveness and environmental protection.

Keywords: rubber fiber; calcareous sand; permeability coefficient; input energy; relative breakage rate

1. Introduction

Calcareous sand is a sediment formed through a series of intricate physical, biological, and chemical reactions involving the remains of marine organisms; Its primary component is calcium carbonate [1]. In contemporary marine engineering construction, calcareous sand plays a crucial role as a vital material in the development of civil engineering infrastructure, particularly in the construction of reefs and islands [2–6] (Figure 1).

Figure 1. Hydraulic reclamation.

Due to the inherent characteristics of calcareous sand, such as its high porosity, pronounced permeability, susceptibility to breakage, and irregular particle shape [7,8], the engineering mechanical properties of calcareous sand exhibit notable distinctions compared to those of conventional natural sands. In recent years, a multitude of researchers have conducted comprehensive experimental investigations encompassing diverse facets of calcareous sand. In terms of the permeability of calcareous sand, pore characteristics are one of the significant factors influencing its permeability. Based on the constant head permeability experiments, Wang et al. [9] discovered that the permeability coefficient of calcareous sand exhibited a linear relationship with 10^e (where e is the void ratio). Using a three-dimensional X-ray microanalysis device, Fan et al. [10] observed the three-dimensional pore characteristics of calcareous sand with different particle sizes, revealing that the quantity and volume of pores in calcareous sand significantly impact its permeability. Moreover, due to the irregular nature of calcareous sand particles, the particle size and shape also exert an influence on permeability [11]. By using the homemade Velocity-controlled Pressure-differential Acquisition Flow Apparatus, Wang et al. [12] indicated that irregular particle shapes reduced the permeability coefficient of calcareous sand. In practical engineering applications, calcareous sand is commonly used as a filling material for subsoil, and is thus inevitably subjected to the loads from the overlying structure. Under the action of the load, the pores between calcareous sand particles are compressed, and the particles themselves undergo stress-induced breakage, leading to changes in the particle shape and size. Therefore, the influence of the load on the permeability of calcareous sand cannot be ignored. By utilizing a piece of homemade permeability equipment, Wang et al. [13] analyzed the effect of the load on the permeability coefficient of calcareous sand. The results demonstrated that, with the gradual increase in load, the permeability coefficient of calcareous sand initially decreased rapidly and then decreased slowly. On the other hand, Li et al. [14] found that impact loads can also significantly reduce the permeability of calcareous sand, and many studies have shown that particle breakage has an impact on permeability [13], indicating a correlation between these two factors. Compared to quartz sands, calcareous sand is more prone to particle breakage [15]. The breakage of larger particles will lead to a decline in the foundation settlement and bearing capacity, posing instability factors and risks to engineering projects [16,17]. Moreover, the increase in the density and confining pressure will also lead to a higher degree of breakage in calcareous sand [18], while the non-uniform calcareous sand particles exhibit significantly stronger deformation resistance and compressive strength [19,20]. At present, most of the studies on the breakage characteristics of calcareous sand focus on particle breakage patterns and the inducing factors such as stress and strain. When it comes to the breakage mechanism of individual calcareous sand particles, the individual calcareous sand particle breakage can be classified into three modes: the primary splitting, the successive breaking, and the severe disintegration [21]. Through the triaxial shear tests of calcareous sand, it has been explained that the shear is the triggering factor for the particle breakage of calcareous sand, and it continues to occur with increasing shear strain under fully plastic conditions after reaching the peak deviator stress [22,23]. From the ring shear tests of calcareous sand, it can be found that the breakage rate of calcareous sand far exceeds the triaxial test; as long as the shear strain is present, the fracture will occur [24,25]. Therefore, both stress and strain can induce particle breakage in calcareous sand. In order to better analyze the particle breakage of calcareous sand and achieve a comprehensive understanding of calcareous sand breakage, some researchers have considered introducing the concept of input energy. Studies have found that the relationship between the breakage rate of the calcareous sand and input energy is independent of the loading conditions, confining pressure, and density [26,27]. Thus, the breakage trend in calcareous sand can be accurately predicted by studying the relationship between the input energy and the breakage rate.

When pure calcareous sand is used as a fill material for island reef construction, it is easily susceptible to groundwater and microbial erosion. Due to its porous nature and significant water absorption capacity, its mechanical properties undergo significant changes, thereby posing potential risks to engineering projects. However, rubber, being a hydrophobic material, exhibits long-term stability in its mechanical properties, as it has minimal reactivity with groundwater. And rubber has advantages in engineering applications because it can reinforce the sand while maintaining its lightweight capabilities, and the waste tire rubber particles, with their low density, lightweight nature, good compressibility, durability, and slow aging in the absence of light [28,29], make it suitable as a new lightweight composite fill material, widely used in the field of geotechnical engineering, including roadbeds [30,31], slopes, concrete [32,33], and retaining walls [34–36]. As the number of private vehicles grows exponentially, countless tires are discarded worldwide each year (Figure 2). The inadequate disposal of these waste tires can cause significant environmental pollution, and direct disposal will also lead to the waste of resources. Therefore, its comprehensive utilization would be beneficial for the development of social economy and ecological preservation [37]. Through consolidation tests, direct shear tests, and triaxial tests, many researchers have investigated the mixing performance of rubber and sand, and concluded that the presence of rubber effectively inhibits the particle breakage behavior in the sand [38,39], suppresses particle reorganization and rearrangement, and delays the shear dilation behavior of the soil [40–42]. In addition, it is of great significance that we incorporate rubber with calcareous sand as a blended foundation fill material in marine engineering construction for the development of marine engineering and environmental conservation. Research on the incorporation of rubber in calcareous sand can provide valuable insights into future reclamation projects of reefs and ground reinforcement projects.

Most of the previous studies focused on the combination of siliceous sand and rubber, and there were few studies on the deformation, permeability, and particle breakage characteristics of the rubber fiber and calcareous sand mixture. To further investigate the influence of rubber fiber on the engineering mechanical properties of calcareous sand, an SLB-1-type stress–strain-controlled triaxial shear permeability test apparatus was employed in this study. The rubber-modified calcareous sand samples with different rubber fiber contents were subjected to triaxial consolidated–drained shear tests (CD), triaxial permeability experiments, and particle analysis tests before and after triaxial compression. The effects of the rubber fiber content on the rubber-calcareous sand's strength, permeability, and breakage characteristics were analyzed. The goal of this study is to provide a reference for future projects to reclaim reefs and subsequent theoretical research.

Figure 2. Stacking of waste tires.

2. Materials and Methods

2.1. Test Materials

The test calcareous sand was taken from the South China Sea. It was an uncemented loose body of particles with extremely irregular surfaces, different sizes, white color, rich internal pores, and different particle shapes such as sheets, spindles, and branches. The rubber fiber was black granular debris obtained by mechanical crushing. The test parameters of the two materials are shown in Table 1, and the particle grading curves obtained after the particle analysis test are shown in Figure 3.

Table 1. Physical parameters of calcareous sand and rubber particles for testing.

Material	G_s	ρ_{dmax}/(gcm^{-3})	ρ_{dmin}/(gcm^{-3})	C_u	C_c	D_{50}/(mm)
calcareous sand	2.71	1.52	1.20	4.71	0.73	1.23
rubber fiber	1.05	0.72	0.44	3.06	1.23	0.46

Figure 3. Particle grading curves of rubber fiber and calcareous sand.

The percentage of rubber fiber content in rubber-modified calcareous sand was expressed by the rubber mass ratio ($RM = m_{rubber}/m_{total}$), where m_{rubber} is the mass of rubber and m_{total} is the total mass of rubber-modified calcareous sand. Since the relative density (D_r) has a great influence on the mechanical properties of the calcareous sand, to ensure the comparability of the test results, D_r was set to 70% to control the loading density of the test, and the calculation formula was as follows:

$$D_{\mathrm{r}} = \frac{\rho_{\mathrm{dmax}}(\rho_{\mathrm{c}} - \rho_{\mathrm{dmin}})}{\rho_{\mathrm{c}}(\rho_{\mathrm{dmax}} - \rho_{\mathrm{dmin}})} \quad (1)$$

$$m = \rho_c V \quad (2)$$

$$m_r = mRM \quad (3)$$

$$m_s = m(1 - RM) \quad (4)$$

where D_r is the relative density of the sample; ρ_{dmax} is the maximum dry density of the sample; and $\boldsymbol{\rho}_{\mathbf{dmin}}$ is the minimum dry density of the sample. ρ_c is the control density of the sample; V is the sample volume; m is the total mass; RM is the proportion of rubber content in the sample; m_r is the mass of rubber; and m_s is the mass of calcareous sand.

The rubber-modified calcareous sand samples for different RMs are shown in Figure 4. The maximum, minimum, and D_r = 70% dry densities of the samples for different RM values are provided in Table 2. The variation in the dry density of the sample with increasing RM is shown in Figure 5. The dry densities of the samples in different states gradually decreased with increasing RM, which showed a nearly linear trend. To prevent uncertainty in the experimental results caused by the salt and binders on the surface of the calcareous sand, the samples were cleaned until the filtrate was transparent and the suspended matter is not obvious and then dried before testing began.

Figure 4. Samples of rubber–sand particle mixtures with different RM.

Table 2. Dry densities of rubber-modified calcareous sand with different mix proportions.

RM/%	ρ_{dmax}/(gcm^{-3})	$\rho_d(D_r = 70\%)$/(gcm^{-3})	ρ_{dmin}/(gcm^{-3})
0	1.52	1.41	1.20
5	1.46	1.33	1.10
10	1.36	1.23	1.00
15	1.29	1.16	0.93
20	1.19	1.07	0.86

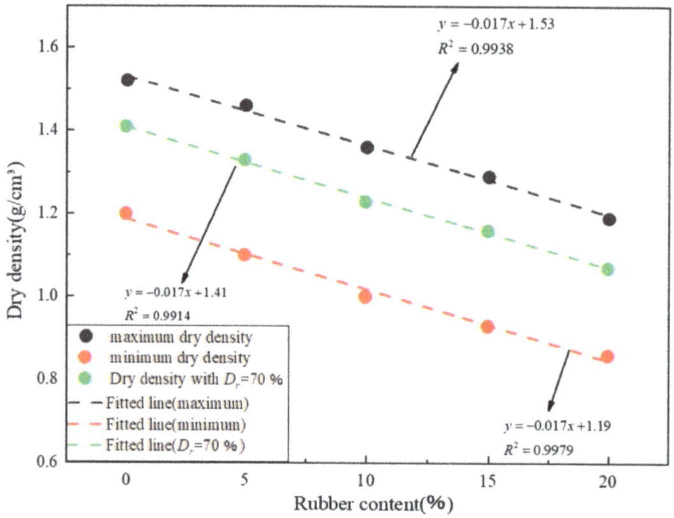

Figure 5. The dry density of the sample changed with increasing content of rubber fiber.

2.2. Experimental Apparatus and Methods

2.2.1. Experimental Apparatus

An SLB-1 stress–strain-controlled shear penetration tester (Figure 6), which was composed of a pressure controller, pressure chamber, and data acquisition system, was utilized to measure the shear strength and deformation characteristics of the calcareous sand. The maximum axial pressure was 20 kN, the maximum circumference range was 2 MPa, and the maximum osmotic pressure was 1 MPa.

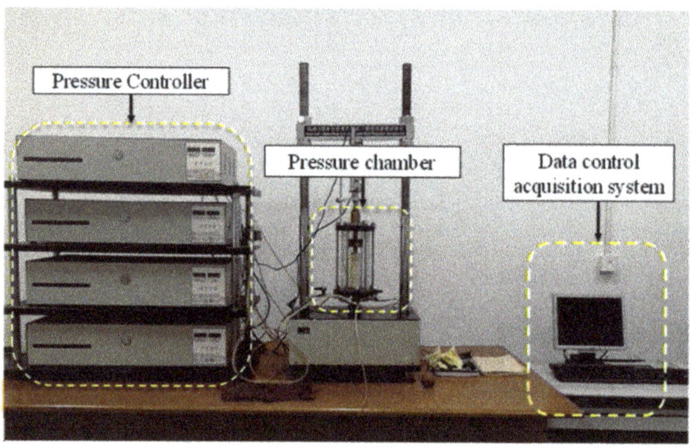

Figure 6. SLB-1-type stress–strain-controlled triaxial shear permeability apparatus.

2.2.2. Test Methods and Procedures

To explore the influence of different RM values on the mechanical properties of rubber-modified calcareous sand, samples with RM values of 0%, 5%, 10%, 15%, and 20% were used to carry out the triaxial consolidation drained shear test, triaxial constant head permeability test, and triaxial compression breakage test. According to the formula for relative density, the quality of each group of samples was calculated and weighed. To ensure the uniformity of each sample, the sample was divided into three equal parts, the stratification method was used to make the sample, and a small hammer was used to prevent particle breakage. After a sample was prepared, back pressure was used to saturate it to 95%.

1. CD test: Rubber-modified calcareous sand with different RM values was subjected to triaxial consolidation drainage shear tests at three different confining pressures (100 kPa, 200 kPa, and 300 kPa). The shear rate was set to 1 mm/min and the test was terminated when the axial strain of the sample reached 20%.
2. Triaxial constant head permeability test: Rubber-modified calcareous sand with different RM was subjected to the triaxial constant head permeability test at six different deviatoric stresses (0 kPa, 500 kPa, 1000 kPa, 1500 kPa, 2000 kPa, and 2500 kPa). The confining pressure of the test was set to 1200 kPa, and the osmotic pressure was set to 10 kPa. During the test, when the axial pressure reached the target, the drainage change of the sample was observed until the drainage volume is stable, and the permeability test was started. When the seepage velocity was stable, the test was stopped. Due to the effect of the osmotic pressure, the internal particles of the sample were transported, thus affecting the structure of the sample, which caused a certain degree of error in the test results. Therefore, every time that a set of tests was completed, the next set of tests was conducted using a new sample.
3. Triaxial compression breakage test: Rubber-modified calcareous sand with different RM was subjected to the triaxial compression breakage test at six different deviatoric stresses (0 kPa, 500 kPa, 1000 kPa, 1500 kPa, 2000 kPa, and 2500 kPa), and the confining pressure and axial pressure were set the same as those in the penetration test. After the compression test, the samples were dried, weighed, and then subjected to particle sieve analysis.

The specific test types, loading conditions, and boundary conditions are shown in Table 3.

Table 3. Test procedure.

Types Of Tests	Confining Pressures	Deviatoric Stresses	Boundary Conditions
CD test	100 kPa, 200 kPa, and 300 kPa	Measured by experiment	Axial strain reached 20%
Triaxial constant head permeability test	1200 kPa	0 kPa, 500 kPa, 1000 kPa, 1500 kPa, 2000 kPa, and 2500 kPa	Seepage stable
Triaxial compression breakage test	1200 kPa	0 kPa, 500 kPa, 1000 kPa, 1500 kPa, 2000 kPa, and 2500 kPa	Deviatoric stresses reach the target value

3. Results and Discussion

3.1. Analysis of the CD Test Results

3.1.1. Stress–Axial Strain Curves

After processing the CD test data of the rubber-modified calcareous sand with different RMs, the deviatoric stress–axial strain curves for different confining pressures were obtained as shown in Figure 7. At the same confining pressure, the initial modulus of the stress–strain curve of each sample decreased with increasing RM, and the curve showed the more obvious characteristics of hardening. The peak stress achieved by the sample decreased with increasing RM, and the axial strain corresponding to the peak stress increased with increasing RM. After the toughness of the low RM sample is weak and reaches the maximum strength, the deviatoric stress decreases rapidly with the increase in strain. For the sample with more than 10% RM, the toughness is obviously enhanced, and the decrease rate of the deviatoric stress becomes slower. The trends were similar to those obtained by

Anvari et al. [43] in CD tests of mixtures of rubber particles with quartz sand at different mix ratios.

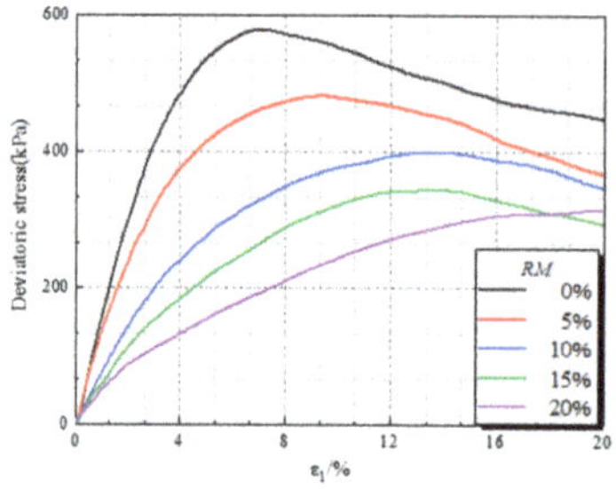

(**a**) Confining pressure 100 kPa

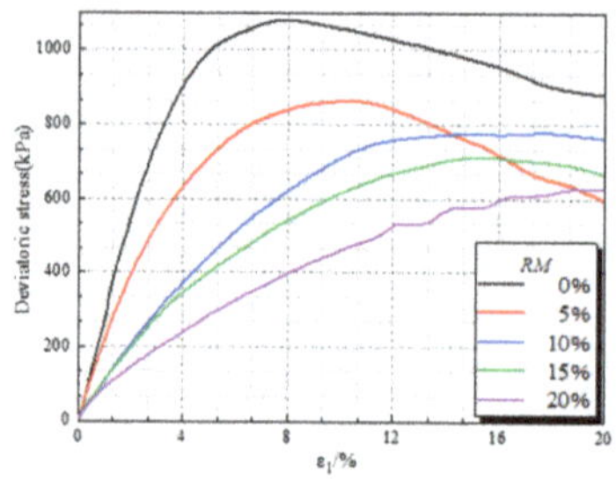

(**b**) Confining pressure 200 kPa

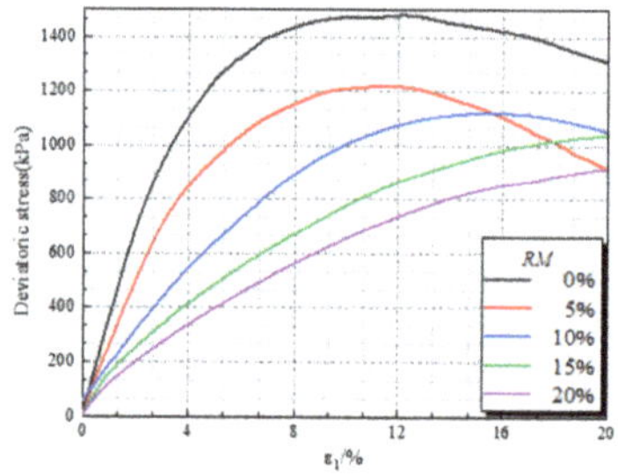

(**c**) Confining pressure 300 kPa

Figure 7. The stress–strain curves of rubber-modified calcareous sand at different *RM*mix ratios for different confining pressures.

For samples with the same RM value, stress–strain curves were compared at different confining pressures. As the confining pressure increased, the peak stress and initial modulus of the samples also increased. The main reason for this was that the higher confining pressure inhibited the movement and breakage of particles and reduced the expansion of the sample volume, thus increasing the shear strength of the sample.

3.1.2. Volumetric Strain–Axial Strain Curves

The characteristics of the volume strain–axial strain of rubber-modified calcareous sand samples with different RMs at different confining pressures are shown in Figure 8, where A represents the shear-contraction stage, B is the stable stage (with the increase in axial strain, the volume of the sample almost does not change), and C is the shear-dilatation stage.

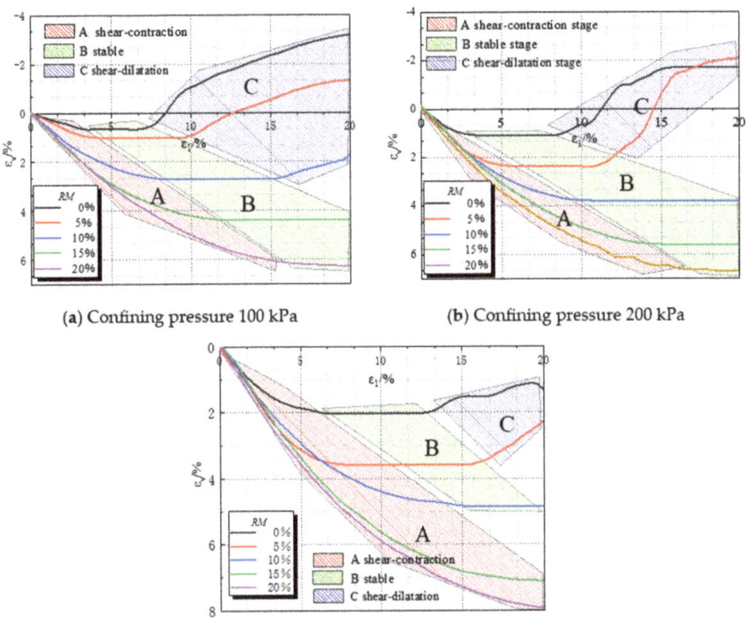

(a) Confining pressure 100 kPa (b) Confining pressure 200 kPa

(c) Confining pressure 300 kPa

Figure 8. Volumetric strain-axial strain curves of rubber-modified calcareous sand at different mix ratios for different confining pressures.

At different confining pressures, the shear-contraction features of the rubber calcium sand were enhanced with increasing RM, and the shear-dilatation features weakened. An RM value of 10% could be used as a critical point. When the sample RM was less than 10%, the sample mainly underwent shear dilatation, and, when the sample RM was more than 10%, the sample was dominated by shear contraction.

The changes in the areas of different regions under different confining pressures in Figure 8 showed that the area of region C decreased obviously, and the area of region A increased, with increasing confining pressure. These results showed that, with increasing confining pressure, samples with different RM values showed the characteristics of weakened shear-dilatation and strengthening shear-contraction effects.

3.1.3. Analysis of the Shear Mechanism of Rubber-Modified Calcareous Sand

Calcareous sand was incompressible and undeformed, but, under high pressure, the extrusion between the particles caused the initially angular particles to be broken into multiple fragments; the particles to become rounder, and the rubber particles had small moduli and a strong ability to undergo deformation (mainly elastic deformation) [44]. Therefore, the shear characteristics of the rubber-modified calcareous sand were closely related to its RM. When the sample had a small RM, the force transmission mechanism under the action of external force was mainly the rigid contact between the particles of calcareous sand, and the shear compression characteristics of the sample were similar to those of unmodified calcareous sand. Before compression, the rubber-modified calcareous sand was very porous and in a loose state. When the axial stress was applied, the pores between the particles decreased rapidly, and the fine particles produced by particle grinding filled the pores of the large particles, making the sample more compact. The stress–strain curve of the sample showed a trend of approximate linear change with continuously increasing stress, and the volume of the sample decreased, which was characteristic of shear contraction. As the stress increased further, some large particles appeared to have become more spherical, which led to a dislocation movement between particles, and the interlocking friction between particles decreased. The rising slope of the stress–strain curve decreased rapidly, the axial strain of the sample continued to increase, the volume of the sample was no longer reduced, and the stress–strain curve of the sample showed characteristics of softening. When the axial stress increased to a certain value, the sample was completely destroyed, and the sample underwent shear dilatation and produced an inclined shear band, and the stress–strain curve showed a downward trend. When the sample had a high RM value, the force transmission mechanism of rubber-modified calcareous sand under the action of external force included rigid contacts between calcareous sand particles, nonrigid contacts between calcareous sand, and rubber and nonrigid contacts between rubber particles; As the sample RM increases, the proportion of the latter two non-rigid contact modes gradually increased. Due to the elastic deformation of the rubber fibers, the smoothing and breakage of calcareous sand particles caused by the stress concentration were inhibited. Additionally, the dislocations between particles were buffered. As a result, the deformability of the specimen was enhanced, and the stiffness of the specimen was reduced. As shown in Figure 9, it was difficult for a shear band to form, and the hardening features of the stress and strain curves of the sample were more obvious. At the same time, during the shear compression process, due to the increase in stress, the rubber fibers were tightly compressed, and they deformed and fully filled the pores of the calcareous sand particles, resulting in a reduction in the pore ratio of the sample. Thus, samples with large RM values exhibited monotonic shear contraction characteristics.

Figure 9. Comparison diagram of sample failure under different RM.

3.2. *Triaxial Permeability Test Results and Analysis*

In the conventional triaxial shear permeability test, the permeability coefficient is calculated from the variation in Darcy's law:

$$k = \frac{\Delta Q \times h}{102 \times \Delta P \times A \times \Delta t} \quad (5)$$

where ΔQ is the total permeation within the time interval Δt; h is the penetration path, which is the height of the sample during penetration; ΔP is the osmotic pressure (1 kPa osmotic pressure equivalent to a 102 mm head); A is the cross-sectional area of the sample during penetration; Δt is the time interval of penetration; and k is the permeability coefficient.

3.2.1. Analysis of the Influence of Load on the Permeability Coefficient of Calcareous Sand

The permeability coefficient–deviatoric stress curves of rubber-modified calcareous sand with different RMs are shown in Figure 10. Figure 10 shows that, with increasing deviatoric stress, the permeability coefficients of the samples with different RMs decreased, and the rate of decrease gradually decreased with increasing deviatoric stress. Each curve could be divided into two stages: rapid reduction (A–B), and the slow decrease and tendency toward stabilization (B–C); 1000 kPa was the boundary between the two stages. In stage A–B, there were multiple pores, calcareous sand particles with irregular shapes, and pores that were extremely sensitive to the change in load. Thus, the pores of rubber-modified calcareous sand were rapidly compressed and reduced by loading, and the water flow was blocked, resulting in a significant decrease in the permeability coefficient in this stage. In the B–C stage, with the increase in deviatoric stress, close contact developed between the particles of the rubber-modified calcareous sand. The particles of calcareous sand did not easily deform. When the original pore structure decreased under deviatoric stress, further compression was difficult. The sample pores were reduced by particle fracture, but the breakage amplitude of the particles was limited. Thus, the slope of the B–C stage of the curve slowly decreased.

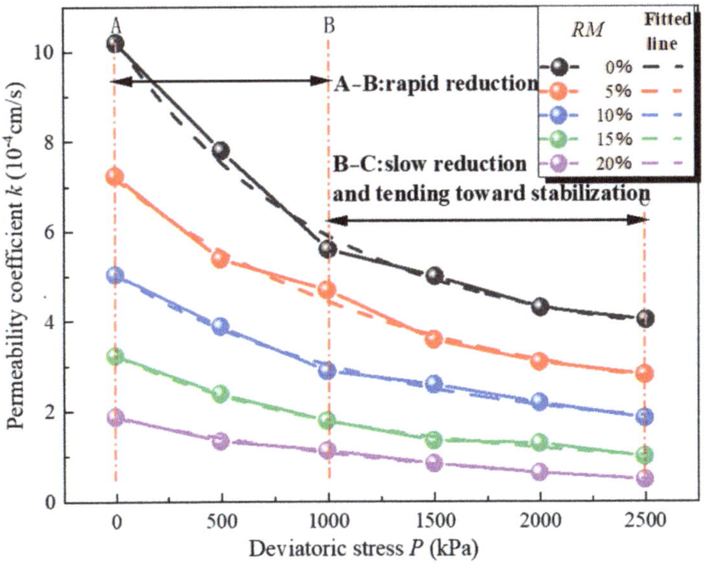

Figure 10. The permeability coefficient–deviatoric stress curves of rubber-modified calcareous sand with different RM values.

Figure 10 showed that, during the whole process of deviatoric stress application, the permeability coefficient variation of samples with a low RM was much larger than that of samples with a high RM. For example, the permeability coefficient of pure calcareous sand samples changed 4.45 times more than that of samples with an RM of 20% and even more significantly in stage A–B, which changed 6.13 times more.

In addition, the permeability coefficient and the deviatoric stress of samples with different RMs could be well-fitted by exponential functions, which were similar to the trends of the experimental curves obtained by Wang Shuai et al. [13] when exploring the influence of the load on the permeability coefficient of calcareous sand. The test data were fitted by Equation (6), and the fitting equations for the samples with different RMs are shown in Table 4.

$$k = \left[A \times e^{(-P/B)} + C\right] \times 10^{-4} \tag{6}$$

where k is the permeability coefficient; p is the deviatoric stress; and $A,\ B,\ C$ are the curve fitting parameters related to the characteristics of rubber-modified calcareous sand.

Table 4. K-P fitted equations for rubber-modified calcareous sand.

RM/%	Fitted Equations	Correlation Coefficients
0	$k = 6.72e^{(-P/976.25)} + 3.45$	$R^2 = 0.9939$
5	$k = 5.23e^{(-P/1339.95)} + 1.98$	$R^2 = 0.9926$
10	$k = 3.60e^{(-P/1223.16)} + 1.45$	$R^2 = 0.9948$
15	$k = 2.48e^{(-P/1148.35)} + 0.77$	$R^2 = 0.9948$
20	$k = 1.78e^{(-P/1740.33)} + 0.09$	$R^2 = 0.9934$

The curve fitting results for each RM are shown in Table 4.

When p tended to infinity, the permeability coefficient k in Equation (6) was $C \times 10^{-4}$. There was a good linear relationship between C and RM, and the fitting formula was as follows:

$$C = (-0.16W + 3.13) \times 10^{-4} \quad (7)$$

The linear correlation between parameter C and complex W reached 0.96, as shown in Figure 11. Therefore, the limit minimum value of the permeability coefficient of rubber-modified calcareous sand with different RMs could be well-predicted according to this formula. The 3.13 in the formula represents the ultimate minimum permeability coefficient of pure calcareous sand, which is a theoretical prediction based on the test results.

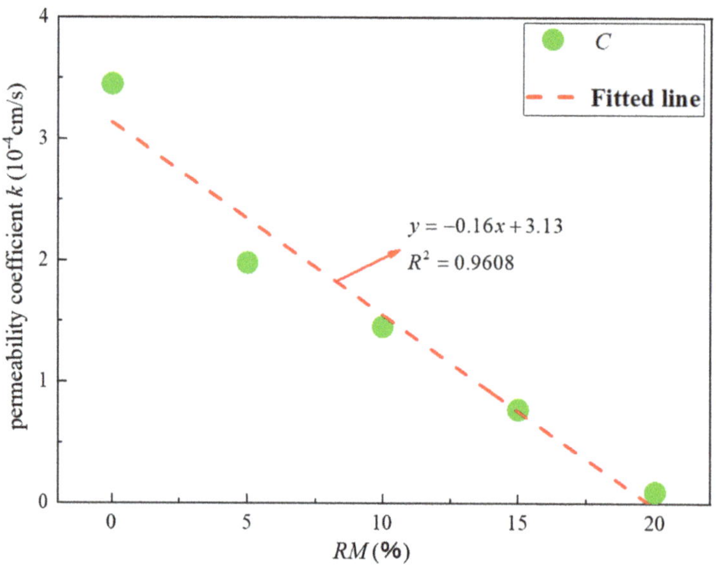

Figure 11. The variation trend of parameter c with RM.

3.2.2. Analysis of the Influence of the RM on the Permeability Coefficient of Rubber-Modified Calcareous Sand

The relationship between the permeability coefficient and the RM of rubber-modified calcareous sand under different deviatoric stresses is shown in Figure 12. According to Figure 8, as the RM increased, the permeability coefficient of the rubber-modified calcareous sand gradually decreased under the same axial pressure. When the deviatoric stress was 0 kPa, the slope of the curve of the graph of the permeability coefficient vs. RM of the sample was 0.413, and when the deviatoric stress increased to 2500 kPa, the slope of the graph of the permeability coefficient vs. RM decreased to 0.178. This showed that, as the axial stress increased, the influence of the RM on the permeability of rubber-modified calcareous sand gradually weakened. This was because, when the deviatoric stress was between 0 kPa and 1000 kPa, the pores inside the sample were not completely compressed, and the rubber fiber was deformed under the stress and filled the pores of the calcareous sand particles. With the increase in the RM, the pores of the calcareous sand particles were more fully filled by the rubber particles, and the permeability of the sample was effectively reduced. When the deviatoric stress was between 1500 and 2500 kPa, the pores between the calcareous sand particles were completely compressed, and the fine particles produced by the friction between the calcareous sand particles filled the original pores to a certain extent. At this time, the decreased permeability coefficient of the sample was limited, and the influence of the RM on the permeability of the sample gradually weakened.

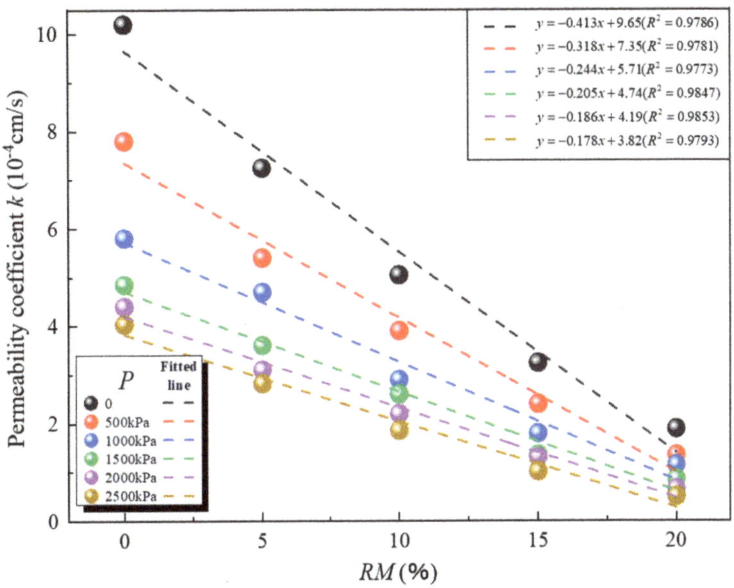

Figure 12. Graphs of permeability coefficient vs. mix ratio for rubber-modified calcareous sand under different deviatoric stresses.

3.3. Triaxial Compression Breakage Test Results and Analysis

Particle breakage is a complex process related to the particle size, particle size distribution, particle shape, void ratio, and particle hardness. The most intuitive manifestation of particle breakage is the difference in the particle gradation curve before and after sample compression. Many scholars take the change at a certain point on the particle grading curve as an indicator of particle breakage. In this study, the breakage model proposed by Hardin [45] was adopted to evaluate the degree of breakage of particles, and the area enclosed by the particle grading curve before compression and the vertical line with particle size D equal to 0.074 mm was defined as the initial breakage potential B_{P0}. It was considered that the particles less than the upper limit of 0.074 mm did not continue to break. The area of the particle gradation curve after compression and breakage and the initial gradation curve and the vertical line with particle size D equal to 0.074 mm were defined as the total breakage B_t (Figure 13). Then, the ratio of the total breakage B_t to the initial breakage potential B_{P0} was defined as the relative breakage rate B_r, as shown below:

$$B_r = \frac{B_t}{B_{p0}} \tag{8}$$

where B_t is the total breakage; B_{P0} is the initial breakage potential; and B_r is the relative breakage rate.

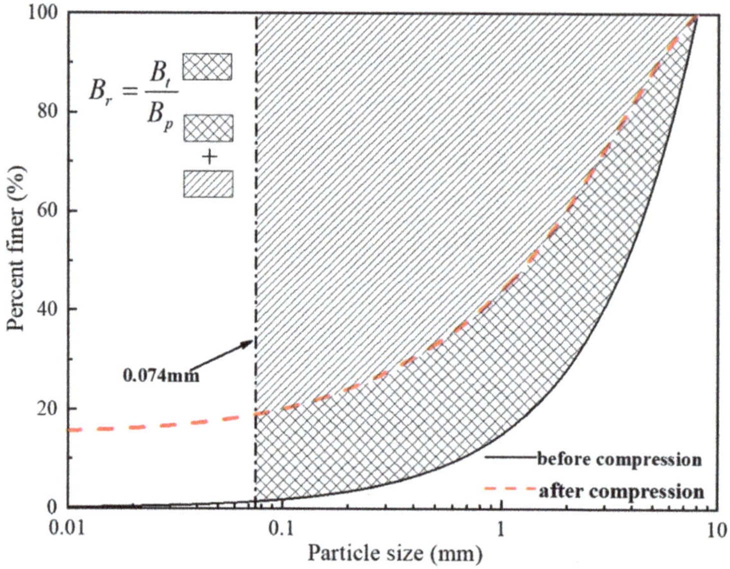

Figure 13. Diagram of the definition of *Br*.

Liu Huabei et al. [27] found that, for the same input energy, the samples with larger particle sizes were more broken than the samples with smaller particle sizes. To explore the effect of different RMs on the particle breakage of calcareous sand, the input energy was analyzed, and the input energy per unit volume of the sample from the beginning of the test to the end could be expressed as follows:

$$W_P = \sum\left[(\sigma_1' - \sigma_3')d\varepsilon_1 + \sigma_3' d\varepsilon_v\right] \tag{9}$$

where W_P is the input energy; σ_1' is the effective major principal stress; σ_3' is the effective minor principal stress; ε_1 is the axial strain increment; and ε_v is the volumetric strain increment.

According to the test results, the relationship between the input energy W_P and the relative breakage rate B_r under different RMs is shown in Figure 14. The degree of breakage of the calcareous sand particles was closely related to the energy absorbed during the stress process and the RM value of the sample. For different RMs, the relationship between W_P and B_r could be fitted by a power function as shown in Equation (10). The curve-fitting formulae for different RMs are shown in Table 5.

$$B_r = a(W_P)^b. \tag{10}$$

where B_r is the relative breakage rate; W_P is the input energy; and a and b are the curve-fitting parameters related to the characteristics of the rubber-modified calcareous sand.

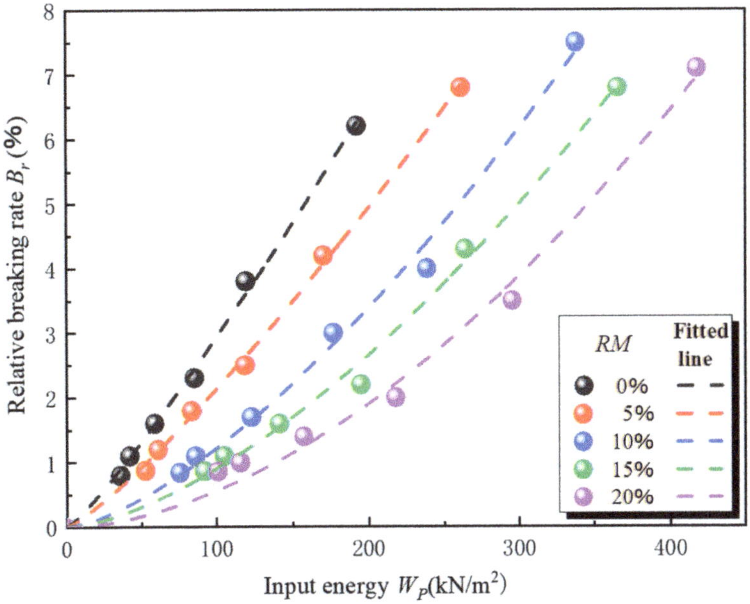

Figure 14. Graphs of relative breakage rate–input energy for rubber-modified calcareous sand with different mix ratios.

Table 5. K-P fitting equations for rubber-modified calcareous sand with different mix ratios.

RM/%	Fitted Equations	Correlation Coefficients
0	$B_r = 1.39 \times 10^{-2} W_P{}^{1.16}$	$R^2 = 0.9965$
5	$B_r = 7.74 \times 10^{-3} W_P{}^{1.22}$	$R^2 = 0.9979$
10	$B_r = 1.13 \times 10^{-3} W_P{}^{1.48}$	$R^2 = 0.9865$
15	$B_r = 6.66 \times 10^{-4} W_P{}^{1.56}$	$R^2 = 0.9929$
20	$B_r = 1.65 \times 10^{-4} W_P{}^{1.76}$	$R^2 = 0.9886$

In the process of energy input, energy is absorbed by the calcareous sand and the rubber. As shown in Figure 14, with increasing RM, the sample B_r-W_P curve gradually became concave. For example, if the sample RM was 20%, then the slope of the curve was small during the initial stage of the energy input. When the input energy exceeded 220kN/m^2, the slope of the curve gradually increased, which indicated that the proportion of energy absorbed by the rubber gradually decreased for a continuous energy input. This was because, during the process of compression, calcareous sand did not deform, and rubber underwent continuous deformation. When the rubber particles compressed and completely filled the pores of the calcareous sand, the contact mode between the particles changed from the rubber particle–calcareous sand particle back to the calcareous sand particle–calcareous sand particle. The change process is shown in Figure 15. As compression continued, the extent of this change continued to grow. When particles contact only calcareous sand particles–calcareous sand particles, the subsequent input energy was mainly absorbed by calcareous sand, which will lead to the breakage of the particles.

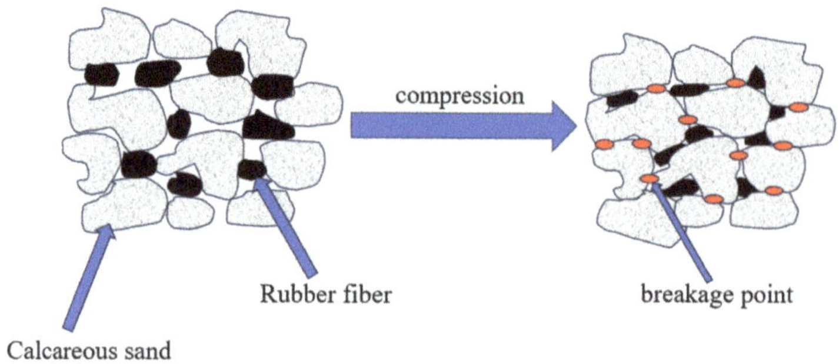

Figure 15. Schematic diagram of particle compression.

In addition, with the same input energy, the relative breakage rates were much smaller for samples with larger RM values than for samples with smaller RM values. There were two main reasons for this behavior; (1) As shown in Figure 16, it can be found that the pores of calcareous sand contain a large number of rubber particles. The presence of rubber fibers reduced the stress concentrations between calcareous sand particles, thus reducing the breakage of particles. (2) The rubber fibers will serve as a skeleton to bear a portion of the load and undergo deformation under the action of the load, thereby absorbing some of the input energy [38,39].

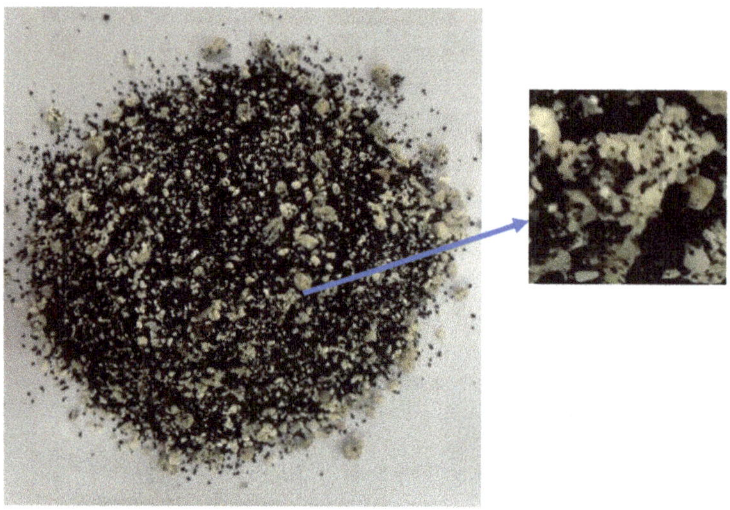

Figure 16. The sample diagram after compression.

4. Conclusions

The paper presents a comprehensive study on the strength, deformation, permeability, and particle breakage characteristics of rubber-modified calcareous sand, investigated through CD tests, triaxial permeability tests, and triaxial compression breakage tests under different rubber content (RM) values and axial stresses. Based on the analysis of the experimental results, the main conclusions are as follows:

1. Increasing rubber content (RM) leads to a decrease in the strength of rubber-modified calcareous sand and an increase in the strain. The stress–strain curve exhibits hardening characteristics, as the modulus of the curve gradually decreases. This behavior is attributed to the presence of rubber, which inhibits shear dilation, enhances the shear

compaction characteristics of calcareous sand, and delays the occurrence of rupture surfaces.

2. Both the rubber content and axial stress have inhibitory effects on the permeability of rubber-modified calcareous sand. The permeability coefficient decreases linearly with increasing RM. On the other hand, with increasing axial stress, the permeability coefficient initially decreases rapidly, then slows down, and eventually converges. The final convergence value is significantly influenced by RM.
3. Rubber has a substantial impact on particle breakage in calcareous sand. It acts in two ways: firstly, it reduces the stress concentration between calcareous sand particles, preventing some particles from undergoing breakage; and, secondly, rubber serves as a skeleton, sharing the load with calcareous sand and undergoing deformation to absorb a portion of the input energy. The relationship between the input energy and the relative breakage rate of calcareous sand particles at different RM levels can be effectively fitted with a power function. Samples with a higher RM show a slow initial increase in the relative breakage rate with increasing input energy, followed by a faster growth rate in the later stage.

Author Contributions: Methodology, Y.C., L.W., and Y.X.; Validation, Y.W.; Formal analysis, H.G.; Investigation, Y.C.; Resources, H.G. and G.Y.; Data curation, Y.W.; Writing—original draft, Y.C.; Writing—review & editing, Y.C., L.W., G.Y., Y.X., L.M., and C.L.; Supervision, L.W. and G.Y.; Project administration, C.L.; Funding acquisition, H.G. and L.M. All authors have read and agreed to the published version of the manuscript.

Funding: This research was funded by the National Natural Science Foundation of China. The grant number were No. 52222110 and No. 52301353.

Data Availability Statement: The raw data supporting the conclusions of this article will be made available by authors on request.

Conflicts of Interest: The authors declare that they have no known competing financial interests or personal relationships that could have appeared to influence the work reported in this paper.

References

1. Wang, X.Z.; Jiao, Y.Y.; Wang, R.; Hu, M.J.; Meng, Q.S.; Tan, F.Y. Engineering characteristics of the calcareous sand in Nansha Islands, South China Sea. *Eng. Geol.* **2011**, *120*, 40–47. [CrossRef]
2. Shen, J.; Xu, D.; Liu, Z.; Wei, H. Effect of particle characteristics stress on the mechanical properties of cement mortar with coral sand. *Constr. Build. Mater.* **2020**, *260*, 119836. [CrossRef]
3. Xu, D.; Chen, W.; Fan, X. Experimental investigation of particle size effect on the self-healing performance of microcapsule for cemented coral sand. *Constr. Build. Mater.* **2020**, *256*, 119343. [CrossRef]
4. Wang, X.; Zhu, C.Q.; Wang, X.Z.; Qin, Y. Study of dilatancy behaviors of calcareous soils in a triaxial test. *Mar. Georesources Geotechnol.* **2018**, *37*, 1057–1070. [CrossRef]
5. Wang, X.; Cui, J.; Zhu, C.Q.; Wu, Y.; Wang, X. Experimental study of the mechanical behavior of calcareous sand under repeated loading-unloading. *Bull. Eng. Geol. Environ.* **2021**, *80*, 3097–3113. [CrossRef]
6. Dai, J.Y.; Yin, S.P.; Hu, C.S. Analysis of key influencing factors of the bond performance between BFRP bars and coral reef and sand concrete. *Constr. Build. Mater.* **2020**, *269*, 121248. [CrossRef]
7. Javdanian, H.; Jafarian, Y. Dynamic shear stiffness and damping ratio of marine calcareous and siliceous sands. *Geo-Mar. Lett.* **2018**, *38*, 315–322. [CrossRef]
8. Wang, X.; Wu, Y.; Cui, J.; Zhu, C.Q.; Wang, X.Z. Shape characteristics of coral sand from the South China Sea. *J. Mar. Sci. Eng.* **2020**, *8*, 803. [CrossRef]
9. Wang, X.Z.; Wang, X.; Chen, J.W.; Wang, R.; Hu, M.J.; Meng, Q.S. Experimental study on permeability characteristics of calcareous soil. *Bull. Eng. Geol. Environ.* **2018**, *77*, 1753–1762. [CrossRef]
10. Fan, Z.; Hu, C.; Zhu, Q.; Jia, Y.; Zuo, D.; Duan, Z. Three-dimensional pore characteristics and permeability properties of calcareous sand with different particle sizes. *Bull. Eng. Geol. Environ.* **2021**, *80*, 2659–2670. [CrossRef]

11. Cui, X.; Zhu, C.; Hu, M.; Wang, R.; Liu, H. Permeability of porous media in coral reefs. *Bull. Eng. Geol. Environ.* **2021**, *80*, 5111–5126. [CrossRef]
12. Wang, Y.; Ren, Y.; Yang, Q. Experimental study on the hydraulic conductivity of calcareous sand in South China Sea. *Mar. Georesources Geotechnol.* **2017**, *35*, 1037–1047. [CrossRef]
13. Wang, S.; Lei, X.W.; Meng, Q.S.; Xu, J.; Xu, Y.; Xie, L. Experimental study on the Influence of Load on Permeability Coefficient of Calcareous San. *IOP Conf. Ser. Earth Environ. Sci.* **2019**, *304*, 052066. [CrossRef]
14. Li, X.; Qiu, Y.; Li, H.; Xu, G.; Xing, H.; Wang, M.; Shi, J. Influence of impact load on permeability of saturated calcareous sand. *Mar. Georesources Geotechnol.* **2024**, *42*, 223–232. [CrossRef]
15. Chen, S.-S.; Zhang, J.-H.; Long, Z.-L.; Kuang, D.-M.; Cai, Y. Effects of particle size on the particle breakage of calcareous sand under impact loadings. *Constr. Build. Mater.* **2022**, *341*, 127809. [CrossRef]
16. Rosa, A.F.; Aragão, F.T.S.; da Motta, L.M.G. Effects of particle size distribution and lithology on the resistance to breakage of ballast materials. *Constr. Build. Mater.* **2020**, *267*, 121015. [CrossRef]
17. Shen, C.-M.; Yu, J.-D.; Liu, S.-H.; Mao, H.-Y. A unified fractional breakage model for granular materials inspired by the crushing tests of dyed gypsum particles. *Constr. Build. Mater.* **2021**, *270*, 121366. [CrossRef]
18. Wu, Y.; Li, N.; Wang, X.; Cui, J.; Chen, Y.; Wu, Y.; Yamamoto, H. Experimental investigation on mechanical behavior and particle crushing of calcareous sand retrieved from South China Sea. *Eng. Geol.* **2020**, *280*, 105932. [CrossRef]
19. Wang, C.; Ding, X.; Xiao, Y.; Peng, Y.; Liu, H. Effects of relative densities on particle breaking behaviour of non-uniform grading coral sand. *Powder Technol.* **2021**, *382*, 524–531. [CrossRef]
20. Xu, D.; Zhang, Z.; Qin, Y.; Yang, Y. Effect of particle size on the failure behavior of cemented coral sand under impact loading. *Soil Dyn. Earthq. Eng.* **2021**, *149*, 106884. [CrossRef]
21. Kuang, D.-M.; Long, Z.-L.; Guo, R.-Q.; Zhao, T.; Wu, K. Experimental and numerical study on the fragmentation mechanism of a single calcareous sand particle under normal compression. *Bull. Eng. Geol. Environ.* **2021**, *80*, 2875–2888. [CrossRef]
22. Wang, G.; Wang, Z.; Ye, Q.; Zha, J. Particle breakage evolution of coral sand using triaxial compression tests. *J. Rock Mech. Geotech. Eng.* **2021**, *13*, 321–334. [CrossRef]
23. Yu, F.W. Influence of particle breakage on behavior of coral sands in triaxial tests. *Int. J. Geomech.* **2019**, *19*, 04019131. [CrossRef]
24. Wei, H.; Zhao, T.; He, J.; Meng, Q.; Wang, X. Evolution of particle breakage for calcareous sands during ring shear tests. *Int. J. Geomech.* **2018**, *18*, 04017153. [CrossRef]
25. Coop, M.R.; Sorensen, K.K.; Freitas, T.B.; Georgoutsos, G. Particle breakage during shearing of a carbonate sand. *Géotechnique* **2004**, *54*, 157–163. [CrossRef]
26. Wang, X.; Liu, J.-Q.; Cui, J.; Wang, X.-Z.; Shen, J.-H.; Zhu, C.-Q. Particle breakage characteristics of a foundation filling material on island-reefs in the South China Sea. *Constr. Build. Mater.* **2021**, *306*, 124690. [CrossRef]
27. Liu, H.B.; Zeng, K.F.; Zou, Y. Particle breakage of calcareous sand and its correlation with input energy. *Int. J. Geomech.* **2020**, *20*, 04019151. [CrossRef]
28. Mohajerani, A.; Kurmus, H.; Conti, D.; Cash, L.; Semcesen, A.; Abdurahman, M.; Rahman, T. Environmental impacts and leachate analysis of waste rubber incorporated in construction and road materials: A review. *Sci. Total Environ.* **2022**, *2022*, 835. [CrossRef]
29. Mashiri, M.; Vinod, J.; Sheikh, M.N.; Tsang, H.-H. Shear strength and dilatancy behaviour of sand–tyre chip mixtures. *Soils Found.* **2015**, *55*, 517–528. [CrossRef]
30. Chompoorat, T.; Jongpradist, P.; Dejdonbomand, C.; Harnnarongchai, W.; Jing, G.; Jamsawang, P. Utilization of Para Rubber Latex and Geopolymer-Stabilized Laterite as Bases and Subbases. *J. Mater. Civ. Eng.* **2025**, *37*, 04025410. [CrossRef]
31. Chompoorat, T.; Sangsai, N.; Tanapalungkorn, W.; Chindasiriphan, P.; Nuaklong, P.; Jongvivatsakul, P.; Likitlersuang, S. Cement-based and alkali-activated controlled low-strength materials made from cup lump rubber for use as road materials. *Road Mater. Pavement Des.* **2025**, *26*, 1151–1171. [CrossRef]
32. Ecemiş, A.S.; Madenci, E.; Karalar, M.; Fayed, S.; Althaqafi, E.; Özkılıç, Y.O. Bending Performance of Reinforced Concrete Beams with Rubber as Form of Fiber from Waste Tires. *Materials* **2024**, *17*, 4958. [CrossRef] [PubMed]
33. Ecemis, A.S.; Madenci, E.; Karalar, M.; Fayed, S.; Althaqafi, E.; OzkillC, Y.O. Shear performance of reinforced concrete beams with rubber as form of fiber from waste tire. *Steel Compos. Struct.* **2024**, *51*, 337.
34. Hazarika, H.; Kohama, E.; Sugano, T. Underwater shake table tests on waterfront structures protected with tire chips cushion. *J. Geotech. Geoenviron. Eng.* **2008**, *134*, 1706–1719. [CrossRef]
35. Lee, J.H.; Salgado, R.; Bernal, A.; Lovell, C.W. Shredded tires and rubber-sand as lightweight backfill. *J. Geotech. Geoenviron. Eng.* **1999**, *125*, 132–141. [CrossRef]
36. Anvari, S.M.; Shooshpasha, I. Influence of size of granulated rubber on bearing capacity of fine-grained sand. *Arab. J. Geosci.* **2016**, *9*, 707. [CrossRef]
37. Mohajerani, A.; Burnett, L.; Smith, J.V.; Markovski, S.; Rodwell, G.; Rahman, M.T.; Kurmus, H.; Mirzababaei, M.; Arulrajah, A.; Horpibulsuk, S.; et al. Recycling waste rubber tyres in construction materials and associated environmental considerations: A review. *Resour. Conserv. Recycl.* **2020**, *155*, 104679. [CrossRef]

38. Liu, L.; Cai, G.; Liu, S. Compression properties and micro-mechanisms of rubber-sand particle mixtures considering grain breakage. *Constr. Build. Mater.* **2018**, *187*, 1061–1072. [CrossRef]
39. Liu, X.; Tian, C.; Lan, H. Laboratory Investigation of the Mechanical Properties of a Rubber–Calcareous Sand Mixture: The Effect of Rubber Content. *Appl. Sci.* **2020**, *10*, 6583. [CrossRef]
40. Dai, B.-B.; Liu, Q.; Mao, X.; Li, P.-Y.; Liang, Z.-Z. A reinterpretation of the mechanical behavior of rubber-sand mixtures in direct shear testing. *Constr. Build. Mater.* **2023**, *363*, 129771. [CrossRef]
41. Cui, M.-J.; Zheng, J.-J.; Dahal, B.K.; Lai, H.-J.; Huang, Z.-F.; Wu, C.-C. Effect of waste rubber particles on the shear behaviour of bio-cemented calcareous sand. *Acta Geotech.* **2021**, *16*, 1429–1439. [CrossRef]
42. Liu, Y.; Liao, X.; Li, L.; Mao, H. Discrete Element Modelling of the Mechanical Behavior of Sand–Rubber Mixtures under True Triaxial Tests. *Materials* **2020**, *13*, 5716. [CrossRef] [PubMed]
43. Anvari, S.M.; Shooshpasha, I.; Kutanaei, S.S. Effect of granulated rubber on shear strength of fine-grained sand. *J. Rock Mech. Geotech. Eng.* **2017**, *9*, 936–944. [CrossRef]
44. Holownia, B.P. Effect of Poisson's ratio on bonded rubber blocks. *J. Strain Anal.* **1972**, *7*, 236–242. [CrossRef]
45. Hardin, B.O. Crushing of soil particles. *J. Geotech. Eng.* **1985**, *111*, 1177–1192. [CrossRef]

Disclaimer/Publisher's Note: The statements, opinions and data contained in all publications are solely those of the individual author(s) and contributor(s) and not of MDPI and/or the editor(s). MDPI and/or the editor(s) disclaim responsibility for any injury to people or property resulting from any ideas, methods, instructions or products referred to in the content.

Journal of
Composites Science

MDPI

Article

Influence of Interface Roughness and Hygrothermal Environment on the Flexural Performance of Prestressed CFRP-Strengthened Cracked Steel Beams

Junhui Li [1,*], Kun Wu [1] and Min Yang [2,*]

1 School of Civil Engineering and Transportation, Foshan University, Foshan 528225, China; 2112361007@stu.fosu.edu.cn
2 Guangdong Hualu Transport Technology Co., Ltd., Guangzhou 510006, China
* Correspondence: jhli@fosu.edu.cn (J.L.); y13670213386@126.com (M.Y.)

Abstract: To meet the strengthening requirements of damaged steel beams in hygrothermal environments, this study conducted four-point bending tests on nine pre-cracked steel beam specimens. The coupled effects of surface roughness, end anchorage, prestressing level of carbon fiber-reinforced polymer (CFRP), and hygrothermal aging on the flexural behavior of the strengthened beams were systematically investigated. Results show that high-grade sandblasting (Sa3) significantly enhances interfacial bond strength through a synergistic "mechanical interlock-adhesion" mechanism, increasing the cracking load of the adhesive layer by 8.2–16.8% compared with Sa2, while partially mitigating the performance degradation caused by hygrothermal aging. The use of end anchorages effectively suppresses CFRP debonding at the beam ends, improving the ultimate load capacity and deformation performance. When a prestress equivalent to 25% of the CFRP's ultimate tensile strength was applied, the load capacity of the strengthened beams further increased by 10.5–19.3%, interfacial cracking was effectively delayed, and the CFRP utilization efficiency reached 96.8–98.5%. Although hygrothermal exposure accelerated interfacial deterioration and reduced the interfacial cracking load, its influence on the ultimate load was relatively limited. These results offer valuable scientific and engineering insights for the design and interface treatment of CFRP-strengthened steel bridges in hygrothermal regions.

Keywords: prestressed CFRP; cracked steel beams; interface roughness; hygrothermal aging; flexural performance

1. Introduction

Steel bridges, characterized by long spans, high load-bearing efficiency, and short construction periods, are widely employed in highway, railway, and urban transportation infrastructure. However, during long-term service, factors such as environmental exposure, increased loads, and functional changes often induce cracking and damage in existing steel structures, necessitating strengthening to ensure operational safety and durability [1]. Consequently, the development of efficient and reliable strengthening techniques for damaged steel bridges has become a critical research focus in the field of civil engineering [2].

Carbon fiber-reinforced polymer (CFRP) has been widely employed for steel structure strengthening due to its high strength-to-weight ratio, corrosion resistance, and excellent fatigue performance [3]. Compared with welding or bolted strengthening, CFRP avoids issues such as heat-affected zone embrittlement or stress concentration around bolt holes, thereby preventing secondary damage to the original structure [4]. However, the effectiveness of CFRP strengthening largely depends on the bonding performance at the CFRP–steel interface [5], and interfacial failure often limits the utilization of CFRP strength to less than 50% for the purely bonding method, hindering its high-strength potential [6].

To improve interfacial performance, physical treatments such as sandblasting and mechanical grinding have been proposed, with sandblasting being widely adopted due to its ease of application and moderate cost [7,8]. Increased surface roughness enhances the "mechanical interlock" effect, significantly improving interfacial resistance to debonding [9]. However, most existing studies focus on the short-term effects of different roughness levels on interfacial performance, while research on the influence of surface roughness on the long-term performance of strengthened beams in service environments remains insufficient. Under high-temperature and high-humidity conditions, the adhesive is prone to aging, leading to interfacial degradation and premature failure [10], but the differences in the ability of bonding interfaces with distinct bonding mechanisms (e.g., "pure bonding" or "mechanical interlocking + bonding") to resist hygrothermal deterioration and maintain long-term stability remain unclear.

In addition, the development of prestressed CFRP technology offers another means to enhance strengthening effectiveness [11]. By applying a pre-tension force, CFRP can partially counteract stresses induced by external loads, reducing the stress intensity factor at crack tips and suppressing crack propagation rates [12,13]. Simultaneously, it optimizes the synergistic behavior between CFRP and steel, significantly improving the utilization efficiency of CFRP materials. It also alleviates the concentration of interfacial shear stress, thereby delaying interfacial debonding [14]. Extensive research and engineering applications have demonstrated that prestressed CFRP strengthening can substantially improve both the static and fatigue performance of steel beams [15–17].

Despite the significant advantages of prestressed CFRP technology in steel bridge strengthening, several key issues need to be clarified when applying this technology to engineering practice. Specifically, the comprehensive influence of laws of critical processes and parameters—such as steel surface treatment, CFRP prestress level, end anchorage, and service environment—on the strengthened system has not been fully addressed in existing studies. Although it is known that hygrothermal environments can accelerate adhesive aging and interfacial degradation, thereby impairing the mechanical performance of the strengthened structure [18–20], the long-term stability of interfaces with different steel surface roughness levels and their effects on the strengthening effectiveness of prestressed CFRP remain unclear. Therefore, it is essential to investigate the impacts of these parameters on the strengthened system to optimize the flexural performance and environmental adaptability of the system.

Against this background, this study focuses on the flexural performance of cracked steel beams strengthened with CFRP and considers four factors: steel surface roughness, end anchor, CFRP prestress, and hygrothermal durations. Nine pre-cracked steel beam specimens were designed, and four-point bending tests were conducted to explore the influence of each parameter on the flexural performance of the strengthened system, providing a scientific basis and practical guidance for the design of CFRP strengthening for damaged steel bridges in hygrothermal regions.

2. Experimental Program

2.1. Materials

For the testing in this work, hot-rolled H-section (HM150 × 100) of Q235B steel beams were chosen as the strengthening objectives, with a total length of 1.4 m. Two types of CFRP plates (Nanjing Hitech Composites Co., Ltd., Nanjing, China) were used to accommodate both non-prestressed and prestressed strengthening schemes. For non-prestressed specimens, plain CFRP plates without end anchors were employed, with a length of 1220 mm, width of 50 mm, and thickness of 2 mm. For prestressed specimens, customized CFRP plates with end anchors were used to facilitate the application of prestress. One end of the CFRP plate was fitted with a fixed anchor for anchorage, while the other end was equipped with a prestressing anchor featuring two longitudinal bolt holes, as shown in Figure 1. The prestressed CFRP plate has a length of 1600 mm, with the same width (50 mm) and thickness (2 mm) as the non-prestressed plate. The adhesive used was Lica-131 CFRP adhesive (Nanjing Hitech Composites Co., Ltd., Nanjing, China), consisting of components A (white) and B (black) mixed at a weight ratio of 2:1 and thoroughly stirred before application.

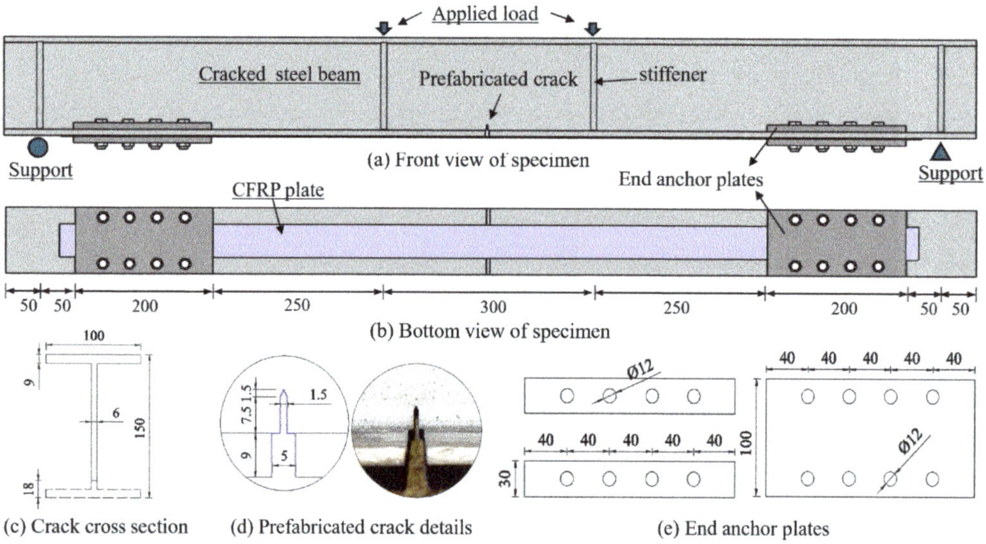

Figure 1. Specimen configuration.

To evaluate material properties, the mechanical behavior of steel beams, adhesive, and CFRP plates was tested using a DDL electronic universal testing machine in accordance with Chinese standards GB/T 2975-2018 [21], GB/T 2567-2021 [22], and GB/T 1447-2005 [23]. The average values of the main mechanical properties are summarized in Table 1.

Table 1. Mechanical properties of materials.

Material	Ultimate Strength (MPa)	Elastic Modulus (GPa)	Yield Strength (MPa)
Q235 steel plate	417.9 ± 7.4	197.3 ± 7.5	258.8 ± 1.8
CFRP plate	2439.5 ± 63	183.2 ± 19.9	—
Lica-131 adhesive	39.2 ± 2.3	5.7 ± 0.6	—

2.2. Specimen Design

To systematically investigate the effects of surface roughness, end anchors, prestress level, and hygrothermal exposure on the flexural performance of damaged steel beams, a total of nine specimens were designed (Table 2). Notably, specimens for evaluating the hygrothermal effect on non-prestressed CFRP-strengthened beams were not included in this study. This design rationale is supported by the authors' previous study [24], which has clearly demonstrated that non-prestressed CFRP-strengthened beams with a pure bonding configuration (without end anchorage) exhibit significant reductions in interfacial cracking load and ultimate load under hygrothermal cycling. In contrast, the present work focuses on the flexural behavior of prestressed CFRP-strengthened beams with end anchors under the same environmental conditions. The exclusion of non-prestressed specimens is therefore reasonable, ensuring a focused evaluation of the combined effects of CFRP prestress and end anchors.

Table 2. Specimen details.

Specimen	Surface Roughness Grade	End Anchors	Prestress Level (F_p/f_u)	Hygrothermal Exposure (Day)
B0	-	without	0	0
B1	Sa2 grade	without	0	0
B1P0		with	0	0
B1P2		with	25%	0
B1P2H		with	25%	90
B2	Sa3 grade	without	0	0
B2P0		with	0	0
B2P2		with	25%	0
B2P2H		with	25%	90

Note: f_u denotes the ultimate tensile strength of the CFRP plate.

The naming of the specimens follows the following principles: Specimen B0, without strengthening, serves as the reference beam; Specimens B1 and B2 correspond to sand-blasting grades Sa2 and Sa3, respectively; P0 and P2 denote the absence of prestress and a prestress level of 25%, respectively; and "H" indicates specimens subjected to 90 days of hygrothermal cycling. According to the GB/T 8923.1-2011 [25], the definitions of Sa2 and Sa3 sandblasting grades are as follows. The Sa2 sandblasting grade refers to thorough blast cleaning, which is characterized by the absence of visible oil, grease, and dirt on the surface when observed without magnification, with almost no scale, rust, coating, and foreign impurities (minimal roughness). The Sa3 sandblasting grade refers to blast cleaning that leaves the steel surface visually clean, which is characterized by the absence of visible oil, grease, and dirt on the surface when observed without magnification, as well as no scale, rust, coating, and foreign impurities; the surface shall have a uniform metallic luster (moderate surface roughness).

To simulate service-induced cracks, initial cracks were prefabricated at the midspan tensile flange and web of the beams. The crack depth was characterized by the relative ratio a/h (where a is the crack depth and h is the beam section height) [18]; in this study, $a/h = 0.12$, corresponding to an initial crack depth of 18 mm. Detailed dimensions of the beams and cracks are shown in Figure 1. To prevent the local premature buckling of the steel beam, 8 stiffeners (10 × 47 × 132 mm) were arranged at the cross sections of the steel beam corresponding to its supports and loading points. Additionally, eight bolt holes were drilled at the ends of the tensile flange for connecting end anchors to secure the CFRP plates. The end anchors, fabricated from Q235 steel plates, have a design dimension of 10 mm (thickness) × 200 mm (length) × 100 mm (width) and are connected to the steel beam using M12 high-strength bolts of grade 8.8. During installation, the bonding adhesive was applied to both sides of the CFRP plate: one side bonded to the steel beam surface, and the other side in contact with the end anchor. The mechanical clamping effect of the high-strength bolts further enhanced the anchorage, ensuring effective composite action between the CFRP plate and the steel beam.

2.3. Specimen Fabrication

Prior to strengthening, the bonding surfaces of the steel beams were sandblasted according to the designed sandblasting grades. The surfaces of the sandblasted steel beams are shown in Figure 2. It can be seen that the specimens with Sa3 sandblasting grade exhibited exposed metallic steel surfaces and noticeable roughness, whereas those with Sa2 sandblasting grade had almost no surface roughness.

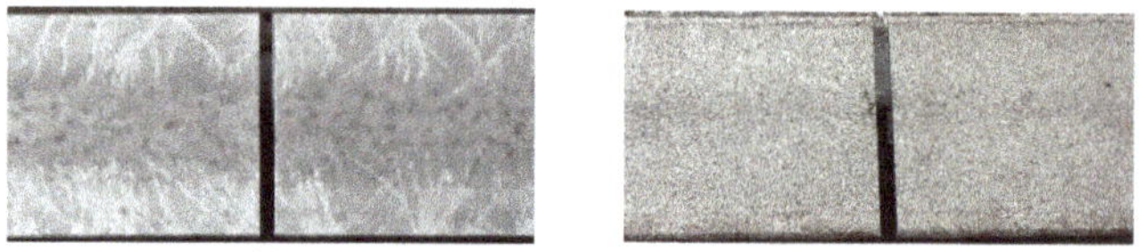

(a) (b)

Figure 2. Surface conditions of steel beams after sandblasting: (**a**) Sa2 grade; (**b**) Sa3 grade.

To facilitate prestress application in the laboratory, a custom prestressing device was developed (Figure 3), comprising a specimen placement section and a prestress application section [18]. The placement section, consisting of vertical plates, compression rods, and adjustable supports allowing horizontal and vertical movement of the steel beam, was used to position and support the beam. The prestress application section, composed of vertical plates, a jack, tie rods, and a stopper plate, connected to the prestressing anchor of the CFRP plate. The jack pushes the baffle, transmitting tensile force through the tie rod to apply prestress.

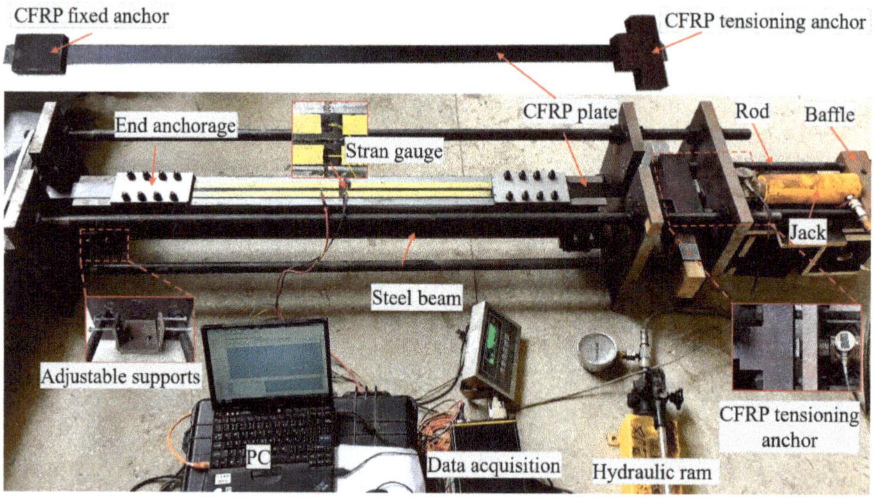

Figure 3. Prestressing setup for CFRP-strengthened steel beams.

The strengthening procedure for prestressed CFRP-strengthened damaged steel beams was as follows: First, strain gauges were attached to the non-bonded surface of the CFRP plate to monitor prestress. The steel beam was placed on the adjustable supports, and the strengthening region was marked. The CFRP plate was installed in the prestressing device and connected via tie rods to the prestressing anchor and stopper plate, with non-strengthened areas and the non-bonded surface of the plate protected by tape. A small initial tensile force was applied via the jack to straighten the CFRP plate, the adjustable supports were used to set a 1 mm gap between the strengthened flange and the CFRP plate and align the strengthening zone. After fixing the steel beam, the jack was released, and the CFRP plate was removed for cleaning the bonding surface with anhydrous ethanol. Next, the Lica-131 adhesive was proportionally measured and mixed with 1% by mass of 1 mm glass beads at low speed to ensure uniform dispersion. The adhesive was evenly applied to the strengthening region of the steel beam, the CFRP plate was reinstalled, and the jack was used to gradually apply the designed prestress, locking the tie rod nuts. End anchor plates were then installed with a pre-tightening torque of 85 N·m, and any excess adhesive was removed. After 24 h of adhesive curing, the tie rod nuts were loosened, the jack unloaded, and the CFRP plate end anchors were cut off, completing the strengthening process.

Accelerated hygrothermal aging of the specimens was conducted using a large-scale hygrothermal cycling system [14,26]. A 3.5% sodium chloride solution was prepared to simulate a marine environment. Each hygrothermal cycle consisted of the following steps: the NaCl solution was heated to 40 °C in a constant-temperature water tank and then pumped into the specimen chamber, immersing the specimens for 10 h. Subsequently, the solution was returned to the water tank, and the specimens were dried in air for 14 h, during which fans at the top of the specimen chamber were activated to accelerate drying. Each cycle lasted 24 h, with a temperature control accuracy of ±1 °C. In this study, the specimens were subjected to 90 hygrothermal cycles.

2.4. Loading and Measurement Scheme

Four-point bending tests were conducted using an SDS500 electro-hydraulic servo universal testing machine. The loading setup and data acquisition system are shown in Figure 4. The span between the loading points was 300 mm, with a clear span of 1300 mm. All bending tests were performed under displacement control at a loading rate of 0.05 mm/s.

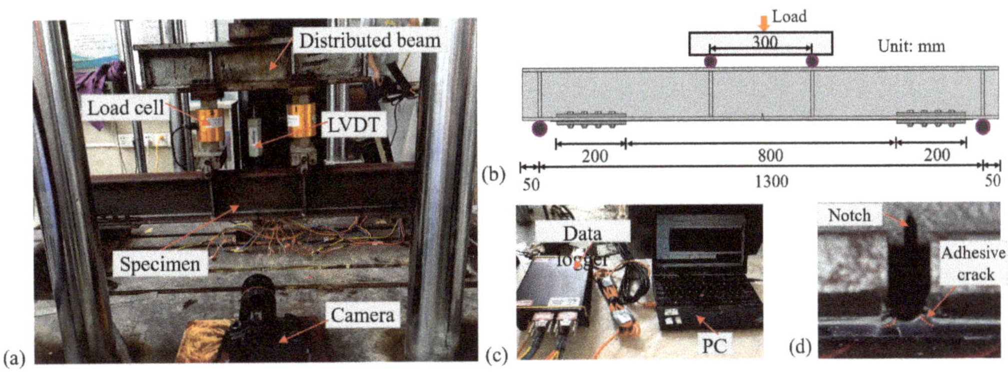

Figure 4. Experimental loading and data acquisition setup: (**a**) loading apparatus; (**b**) schematic of the loading configuration; (**c**) data acquisition system; (**d**) Interface at notch position from camara perspective.

During loading, high-resolution photographs of the interface at midspan cracks were taken to monitor CFRP debonding (Figure 4a,d). The camera was set to automatically capture one image every 2 s, and the time of the first photograph was recorded. By subsequently matching the recorded load data (with corresponding timestamps) from the data acquisition system, the interfacial cracking load at the onset of debonding can be determined. To obtain the yield and cracking loads of the prefabricated crack in the steel beams, two strain gauges were placed at the crack tips (Figure 5); the load at which a crack-tip strain gauge failed was defined as the web cracking load. To capture the strain distribution along the CFRP plate during loading, eleven strain gauges were arranged along its longitudinal axis. Since the interfacial stress is more complex at midspan cracks, the strain gauges were positioned more densely in this region (Figure 5).

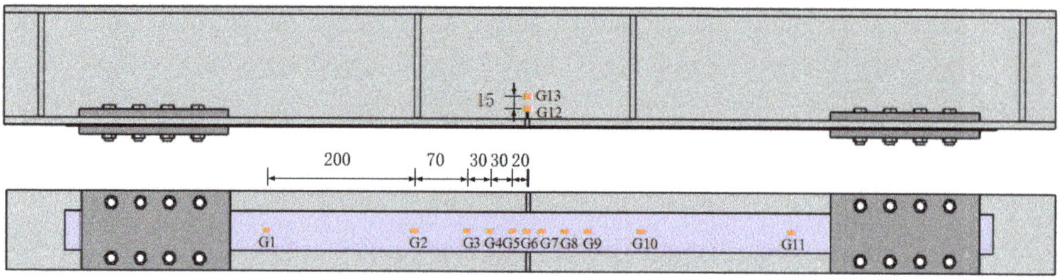

Figure 5. Strain gauge arrangement.

3. Experimental Results and Discussion

3.1. Load–Deflection Response

The load-deflection response of all specimens is presented in Figure 6. The load–deflection curve of the reference beam (B0) exhibited the typical "elastic–plastic–brittle failure" behavior, which reflects the evolution of the initial web cracks. In the elastic stage, the internal stresses did not exceed the steel yield limit, and the load–deflection relationship remained approximately linear. Once the load surpassed the elastic threshold, stress concentration at the crack tip induced plastic deformation, causing a marked increase in deflection rate. Ultimately, at the ultimate load, the crack rapidly propagated through the section, significantly reducing the effective cross-sectional area and leading to sudden brittle failure. This behavior serves as a baseline for evaluating the performance of strengthened specimens.

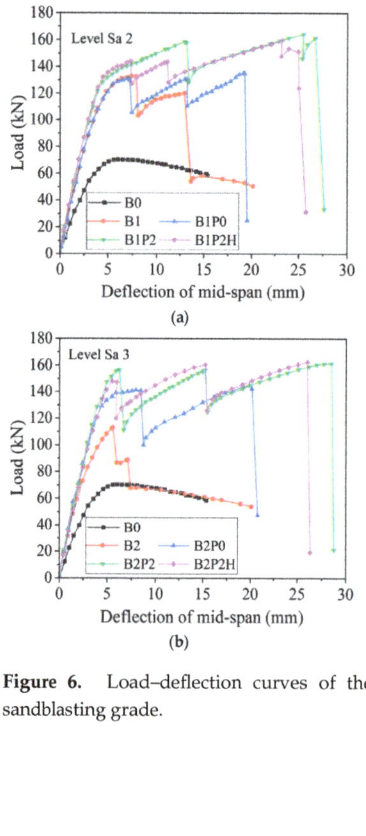

Figure 6. Load–deflection curves of the specimens: (**a**) Sa2 sandblasting grade; (**b**) Sa3 sandblasting grade.

Non-prestressed CFRP-strengthened specimens effectively extended the elastic stage and significantly enhanced stiffness through the synergistic action between the CFRP plate and the steel beam. The key mechanism lies in the CFRP sharing tensile stresses and inhibiting early crack plasticization. The presence of end anchors constrained the slip at the CFRP plate ends, promoting a more uniform interfacial shear stress distribution, thereby delaying adhesive layer cracking and maintaining residual load-carrying capacity after partial CFRP debonding. In contrast, specimens without anchors exhibited load–deflection behavior that rapidly approached that of B0 after debonding. Sandblasting grade also substantially influenced the interfacial bonding mechanism: Sa3 sandblasting formed a "mechanical interlock + adhesion" composite interface, providing superior shear resistance compared with the adhesion-only Sa2 interface. Consequently, Sa3 specimens (B2, B2P0) showed delayed adhesive cracking and debonding. However, due to errors in the experimental arrangement for specimen fabrication and loading, specimen B2 did not receive sufficient curing time for the adhesive layer. Consequently, it exhibited a lower peak load compared to specimen B1. This also indirectly indicates that in the practical application of engineering reinforcement, sufficient curing of the interface requires focused attention.

Prestressed specimens (B1P2, B2P2) demonstrated a more stable linear elastic stage because the CFRP plates were already in tension at the early loading stage, enhancing the synergistic action with the steel beam and slowing the load drop after debonding, thereby improving structural ductility relative to non-prestressed specimens. Hygrothermal cycling (B1P2H, B2P2H) accelerated adhesive aging and interfacial degradation, shortening the elastic stage, advancing debonding, and promoting brittle failure modes (e.g., CFRP rupture at ultimate load in B2P2H). Notably, Sa3 sandblasting partially mitigated the detrimental effects of the hygrothermal environment. The "mechanical interlock" interface exhibited enhanced resistance to degradation, resulting in higher first debonding load and ultimate load for B2P2H compared with B1P2H with Sa2 treatment. This indicates that high-roughness surface treatment can delay the degradation of CFRP-strengthened systems under hygrothermal service conditions.

3.2. Analysis of Flexural Performance Indicators

To compare the effectiveness of different strengthening schemes under various conditions, the loads corresponding to steel beam crack-tip yielding, adhesive layer cracking, web cracking, and ultimate failure are summarized in Table 3. The load at crack-tip yielding was determined using strain gauges attached at the crack tips, while the loads for adhesive layer cracking and web cracking were identified from photographs taken during testing. The ultimate load is defined as the load at final failure of the specimen. Note that data for specimen B1P0 are not included in Table 3 due to strain gauge malfunction during the test.

Table 3. Experimental results of the specimens.

Specimen	Crack-Tip Yield Load (kN)	Adhesive Cracking Load (kN)	Web Cracking Load (kN)	Ultimate Load (kN)	Bending Stiffness (kN/mm)	Crack-Tip Yield Deflection (mm)	Ultimate Deflection (mm)
B0	52.6	—	67.5	70.8	15.7	2.8	6.5
B1	94.5	110.9	117.6	120.3	30.2	3.3	13.8
B1P0	—	114.0	—	135.5	31.3	—	19.4
B1P2	120.0	136.0	149.2	161.5	32.7	3.8	26.8
B1P2H	101.3	120.2	138.5	159.6	33.2	3.1	22.9
B2	74.3	81.3	85.0	89.0	29.8	2.6	7.2
B2P0	83.1	133.1	140.8	145.4	32.7	2.4	20.3
B2P2	113.3	147.1	156.5	161.5	34.1	3.5	28.6
B2P2H	114.6	137.8	148.7	162.7	34.3	3.7	26.1

The adhesive cracking loads of different specimens are presented in Figure 7a. As previously noted, specimen B2 exhibited a lower cracking load due to insufficient adhesive curing. Comparison between non-prestressed and prestressed specimens indicates that the application of prestress significantly increased the adhesive cracking load and effectively delayed adhesive failure. Under the Sa2 sandblasting grade, the prestressed specimen B1P2 showed a 19.3% increase in adhesive cracking load compared with the non-prestressed B1P0, whereas after 90 hygrothermal cycles, B1P2H decreased by 11.6% relative to B1P2. Under the Sa3 sandblasting grade, the prestressed specimen B2P2 exhibited a 10.5% increase over B2P0, and B2P2H decreased by 6.3% after 90 hygrothermal cycles. Moreover, the variation in cracking load for Sa3 specimens was smaller than that for Sa2 specimens. Further comparison between specimens with different surface roughness under the same conditions shows that B2P0 increased the cracking load by 19.1 kN relative to B1P0, B2P2 increased by 11.1 kN relative to B1P2, and B2P2H increased by 17.6 kN relative to B1P2H. These results indicate that higher surface roughness effectively enhances interfacial bond strength and the corresponding adhesive cracking load, and better mitigates the deterioration of the interface under hygrothermal conditions.

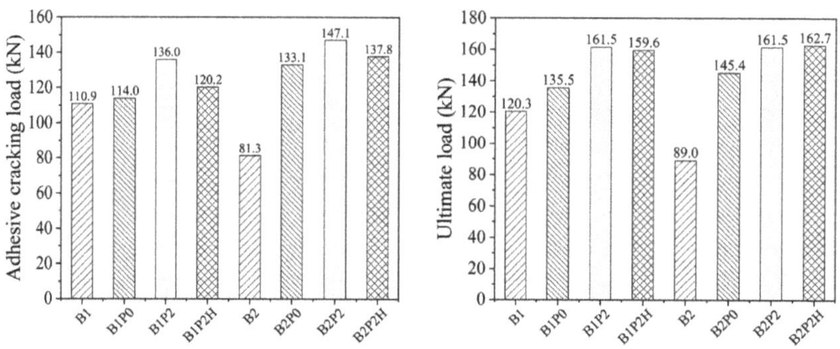

(a) (b)

Figure 7. Characteristic loads of the specimens: (**a**) adhesive cracking load; (**b**) ultimate load.

The ultimate loads of different specimens are shown in Figure 7b. Under the Sa2 sandblasting grade, the specimen B1P0 with end anchors exhibited an ultimate load 12.2 kN higher than the unanchored specimen B1. This improvement is primarily attributed to the end anchors allowing the CFRP plate to continue carrying load after full-length debonding, thereby enhancing the safety reserve of the strengthened steel beam. Meanwhile, the prestressed specimen B1P2 showed an ultimate load increase of 26 kN compared with the non-prestressed B1P0, whereas the difference between B1P2H after 90 hygrothermal cycles and B1P2 was only 1.9 kN. Specimens under the Sa3 sandblasting grade exhibited similar trends (B2 was excluded due to insufficient adhesive curing). The prestressed B2P2 had an ultimate load 16.1 kN higher than the non-prestressed B2P0, with only a 1.2 kN reduction after 90 hygrothermal cycles (B2P2H). Overall, the hygrothermal environment has a significant effect on adhesive cracking load but a relatively minor effect on ultimate load. This indicates that hygrothermal aging primarily accelerates interfacial degradation; however, when end anchors are employed, its impact on the overall load-carrying capacity of the strengthened beams is limited.

In this study, stiffness is defined as the slope of the load–midspan deflection curve prior to the occurrence of interfacial cracks in the steel beam, representing the load-carrying stiffness of the strengthened beam during the elastic stage. The relevant parameters are listed in Table 3. The stiffness of all specimens is shown in Figure 8. Compared with the reference specimen B0, all strengthened specimens exhibited a significant increase in elastic-stage stiffness. Non-prestressed specimens with different surface roughness showed similar stiffness improvements, approximately 91.1%, whereas prestressed specimens exhibited an overall increase of about 112.7%. Notably, the additional contribution of prestress to stiffness was not significant compared with non-prestressed specimens, consistent with previous studies [27,28], indicating that the stiffness of CFRP-strengthened beams primarily depends on the geometry and mechanical properties of the CFRP rather than the prestress level when the elastic modulus of CFRP is comparable to that of steel. Furthermore, comparison between prestressed specimens with and without hygrothermal treatment shows that 90 hygrothermal cycles had a limited effect on stiffness, suggesting that under the experimental conditions of this study, the hygrothermal environment has minimal influence on the elastic-stage performance of CFRP-strengthened steel beams.

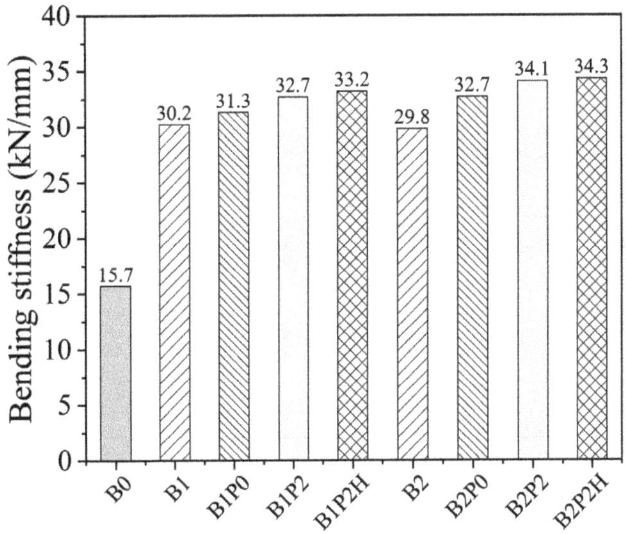

Figure 8. Stiffness of the specimens.

The ultimate deflections of the specimens are shown in Figure 9. Compared with the reference specimen B0, the ultimate deflections of unanchored specimens B1 and B2 increased by 112.3% and 10.8%, respectively. The lower ultimate deflection of B2 is attributed to premature interface debonding caused by insufficient adhesive curing. For unanchored strengthened beams, the CFRP plate tends to propagate debonding rapidly toward the plate ends, eventually resulting in complete end debonding and failure of the strengthening system. In contrast, strengthened beams with end anchors exhibited substantially higher ultimate deflections, regardless of prestress application, with increases ranging from 3.0 to 4.4 times. This indicates that end anchors effectively inhibit plate-end debonding and enhance the ductile reserve of the beams. The hygrothermal environment has a moderate adverse effect on the ultimate deflection of the strengthened beams. For Sa2 sandblasted specimens, the ultimate deflection of B1P2H decreased by 14.6% compared with B1P2, while for Sa3 specimens, B2P2H decreased by 8.7% relative to B2P2. These results demonstrate that higher steel surface roughness helps mitigate deflection degradation under hygrothermal conditions.

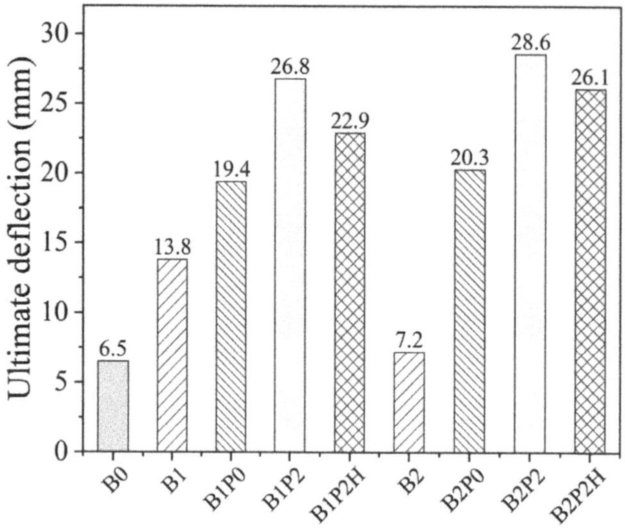

Figure 9. Ultimate deflection of the specimens.

3.3. Load–Strain Analysis

Experimental results indicate that the strain distribution patterns of the CFRP plates are generally consistent across different surface roughness treatments. Figure 10 presents the strain distribution curves of the CFRP plates at various load levels for all specimens. It should be noted that one strain gauge near the end anchor of Specimen B2P2 failed in the later stage of loading; therefore, there is one missing data point in the curve under the failure load. As shown, all specimens exhibited stress concentration at the midspan crack. Under the same load, the CFRP strain gradually decreased from the midspan defect toward the plate ends. With increasing load, the stress concentration became more pronounced: when the load reached the adhesive cracking load at the CFRP–steel interface, the strains on both sides of the defect approached those at the midspan crack. As the load increased further, strains around the defect continued to rise, and the high-stress zone gradually extended toward the plate ends. When interfacial debonding approached the end anchor locations, strains farther from the midspan also increased significantly. However, due to the anchoring effect provided by the end anchors, the specimen can sustain the load continuously. At ultimate load, the CFRP plate was almost fully debonded along its length, and the strain distribution became relatively uniform. For the specimens without end anchors (B1 and B2), due to the long bonding length of the CFRP plates, after the first interfacial debonding occurred, the load dropped suddenly, but the CFRP plates did not debond completely, and the strengthened beams could still maintain their load-carrying capacity. However, when the load increased only slightly afterward, one end of the CFRP plates underwent complete debonding, and the strengthened beams finally lost their load-carrying capacity. As a result, the strain at the ultimate load showed a decreasing trend. In addition, compared with the specimens with end anchors, the high-strain areas of the specimens without end anchors were smaller and only concentrated in the mid-span region. These results demonstrate that end anchors effectively enhance the load-carrying capacity of the specimens and allow the high-strength potential of the CFRP plates to be fully utilized.

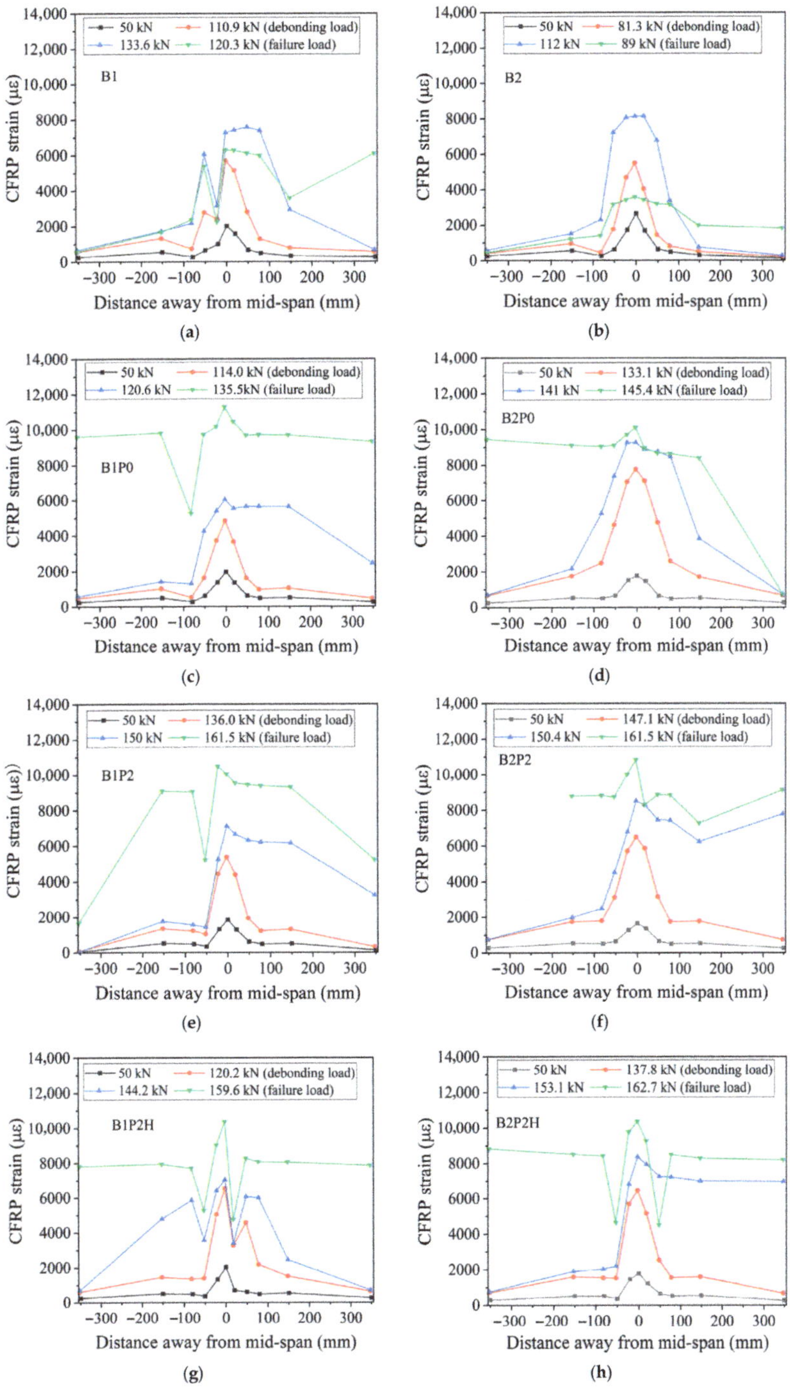

Figure 10. Longitudinal strain distribution of CFRP plates in the strengthened beams. (**a**) specimen B1; (**b**) specimen B2; (**c**) specimen B1P0; (**d**) specimen B2P0; (**e**) specimen B1P2; (**f**) specimen B2P2; (**g**) specimen B1P2H; (**h**) specimen B2P2H.

3.4. Failure Modes

The failure modes of strengthened beams with different surface roughness treatments were generally consistent. Taking the Sa3 sandblasted specimens as an example (Figure 11), all specimens initiated debonding from the midspan prefabricated crack. As the load increased, instantaneous debonding occurred at the adhesive interface, causing a temporary drop in load. This behavior is primarily attributed to the relatively long bonded length of the CFRP plate; after instantaneous debonding, the interfacial cracks had not yet propagated to the plate ends, allowing the strengthened system to continue carrying load. For non-prestressed specimens with end anchors, failure was characterized by the CFRP plate being pulled out from the end anchors. In contrast, prestressed specimens, regardless of environmental exposure, exhibited a mixed failure mode combining CFRP rupture and end-anchor pullout. These results indicate that prestressed strengthening enables CFRP plates to fail in the material rather than solely by interfacial debonding, thereby significantly enhancing the utilization of the CFRP's high-strength capacity at ultimate load.

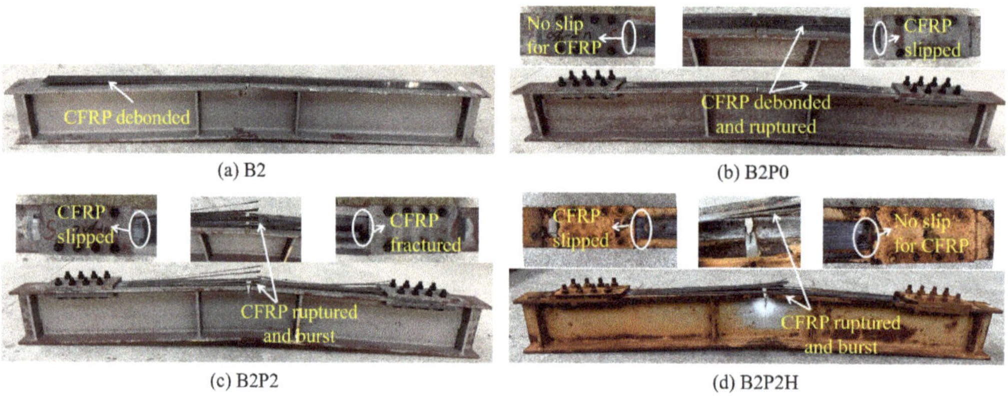

Figure 11. Failure modes of strengthened beams with Sa3 sandblasting grade.

As shown in Table 3, the sandblasting grade has a significant influence on the interfacial debonding behavior of the strengthened beams. Figure 12 presents the local failure modes at the midspan crack: specimens with Sa2 treatment (B1P0, B1P2, B1P2H) primarily exhibited failure at the steel–adhesive interface, whereas specimens with Sa3 treatment (B2P0, B2P2, B2P2H) first experienced cohesive failure within the adhesive layer, followed by debonding at the CFRP–adhesive interface. From the perspective of interfacial formation mechanisms, the CFRP–adhesive interface constitutes a strong phase, whereas the steel–adhesive interface is a weak phase, making the steel/adhesive interface more susceptible to failure. Increasing the roughness of the steel bonding surface promotes the formation of an interlocking dentate structure between the adhesive and steel surface, thereby enhancing bond strength, improving interfacial cracking resistance, and altering the failure mode. After hygrothermal treatment, corrosive media penetrate more easily into the Sa2 interface compared with the higher-roughness Sa3 specimens, resulting in a more pronounced reduction in steel/adhesive bond strength and a decrease in adhesive cracking load. Comparison of the local failure in B1P2H and B2P2H shows rust formation at the bond interface in B1P2H, whereas B2P2H remained free of corrosion, indicating that a superior bonding interface helps resist the ingress of corrosive agents.

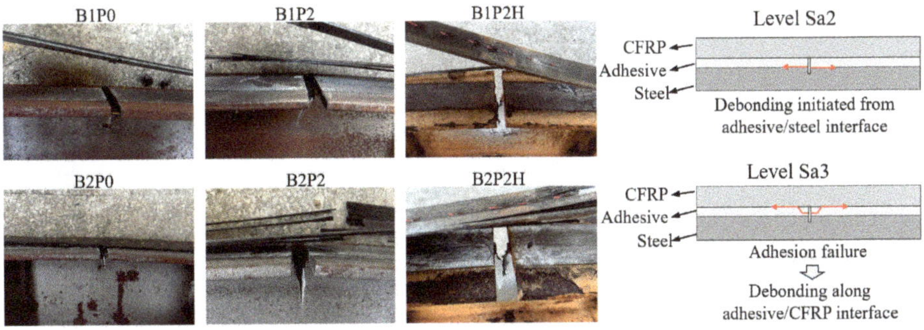

Figure 12. Interfacial debonding characteristics at the crack location of the strengthened beams (the red arrow means debonding propagation path).

3.5. CFRP Strength Utilization

The strength utilization of CFRP plates is an important indicator for evaluating the effectiveness of CFRP in structural strengthening. In this study, the ratio of the midspan strain to the ultimate tensile strain of the CFRP plate (13,316 με, calculated from the data in Table 2) is defined as the debonding utilization and ultimate utilization. For non-prestressed specimens, the initial strain is taken as zero, while for prestressed specimens, it corresponds to the strain after prestress application. The results are summarized in Table 4. It should be noted that the calculation of CFRP ultimate utilization efficiency in Table 4 does not account for the prestress loss of specimens after construction. As confirmed by the authors' previous study [6], specimens prestressed to 40% of the CFRP tensile capacity and anchored at the CFRP ends exhibited a negligible prestress loss of approximately 40 με over 100 days (corresponding to 0.73% of the initial prestress). The loss for specimens with 25% prestress in this study is expected to be even smaller and was therefore neglected. Although this omission may cause a slight overestimation, the results still reliably reflect the strengthening effect of CFRP prestress and end anchors under hygrothermal conditions.

Table 4. Comparison of CFRP plate strength utilization.

	Initial Strain (με)	Debonding Strain (με)	Strain Sum (με)	Debonding Utilization (%)	Ultimate Strain (με)	Strain Sum (με)	Ultimate Utilization (%)
B1	0	5713	5713	42.9	7284	7284	54.7
B1P0	0	6055	6055	45.5	10,305	10,305	77.4
B1P2	2376	5391	7767	58.3	10,509	12,885	96.8
B1P2H	2452	6549	9001	67.6	10,400	12,852	96.5
B2	0	5531	5531	41.5	3176	3176	23.9
B2P0	0	7780	7780	58.4	10,124	10,124	76.0
B2P2	2255	6500	8755	65.7	10,863	13,118	98.5
B2P2H	2273	6470	8743	65.7	10,392	12,665	95.1

In general, the surface roughness of steel has no significant effect on the strength utilization rate of CFRP plates in strengthened beams with end anchors. This is because the final failure mode of all strengthened beams in this study is determined by the end anchors. In the following sections, specimens with Sa3 surface roughness will be taken as an example to discuss the strength utilization rate of CFRP. For the specimen B2, the CFRP utilization at interface debonding was 41.5%, decreasing to 23.9% at complete debonding, due to the abrupt load drop after interface failure, which prevented the load from reaching the debonding threshold. In contrast, specimen B1, with more adequate adhesive curing, was able to sustain a load up to the ultimate capacity even after interface debonding, resulting in a significantly higher ultimate utilization. For specimens with end anchors, such as B2P0, the utilization at interface debonding and ultimate failure reached 58.4% and 76.0%, respectively, indicating that end anchors substantially enhance the ultimate utilization of CFRP plates. Prestressed strengthening further improved material efficiency: for specimen B2P2 with a 25% prestress level, CFRP utilization at interface debonding and ultimate load was 65.7% and 98.5%, representing increases of 12.5% and 29.6% compared with the non-prestressed specimen B2P0. This demonstrates that prestressing not only enhances utilization at interface debonding but also more significantly improves utilization of ultimate strength. After 90 days of hygrothermal cycling, specimen B2P2H exhibited utilizations of 65.7% and 95.1% at interface debonding and ultimate load, respectively. The decrease in utilization of ultimate strength is only 3.4% relative to B2P2, indicating that environmental effects have a limited impact on CFRP strength utilization.

4. Conclusions

This study investigated the flexural performance of CFRP-strengthened steel beams under hygrothermal conditions by conducting four-point bending tests on nine specimens with varying sandblasting grades (Sa2, Sa3), prestress levels (0%, 25% of CFRP ultimate tensile strength), and hygrothermal exposure durations (0 and 90 days). The effects and mechanisms of these variables on the strengthened beams were systematically analyzed, and the main findings are summarized as follows:

(1) End anchors effectively suppressed rapid failure following CFRP debonding, increasing the ductility coefficient by 3.1–3.7 times compared with unanchored specimens. When combined with 25% prestress of the CFRP ultimate tensile strength, the ultimate loads of specimens B1P2 (Sa2 grade) and B2P2 (Sa3 grade) increased by 26 kN and 16.1 kN, respectively, relative to non-prestressed specimens (specimens B1P0 and B2P0). The CFRP strength utilization reached 96.8–98.5%, fully exploiting the high-strength material properties and achieving a balance between structural safety and economic efficiency.

(2) Sa3 sandblasting created a "mechanical interlock + adhesive bonding" composite interface, resulting in adhesive cracking loads 11.1–19.1 kN higher than the purely adhesive Sa2 interface. After 90 days of hygrothermal aging, the cracking load of Sa3 specimens decreased by only 6.3%, considerably less than the 11.6% reduction observed for Sa2, and no interface corrosion occurred. In practice, for controllable steel beam damage, especially in hygrothermal environments, Sa3 grade sandblasting is recommended to enhance interfacial durability.

(3) Although hygrothermal cycling accelerates the degradation of the adhesive cracking load, its influence on ultimate load is minimal, with a maximum difference of only 1.9 kN, indicating that overall structural capacity remains controllable. In engineering applications, combining high-roughness surface treatment with end-anchored prestressed CFRP, together with enhanced interface sealing and protection measures, can further mitigate the adverse effects of hygrothermal exposure on interfacial performance.

Author Contributions: Conceptualization, J.L.; methodology, J.L.; validation, J.L. and M.Y.; formal analysis, M.Y.; investigation, K.W.; resources, M.Y.; data curation, J.L.; writing—original draft preparation, K.W.; writing—review and editing, J.L.; visualization, K.W.; supervision, J.L.; project administration, J.L.; funding acquisition, J.L. All authors have read and agreed to the published version of the manuscript.

Funding: This research was funded by Guangdong Province General Universities Young Innovative Talents Program (Grant No. 2024KQNCX148).

Institutional Review Board Statement: Not applicable.

Data Availability Statement: The original contributions presented in this study are included in the article. Further inquiries can be directed to the corresponding authors.

Conflicts of Interest: Author Min Yang was employed by the company Guangdong Hualu Transport Technology Co., Ltd., Guangzhou. The remaining declare that the research was conducted in the absence of any commercial or financial relationships that could be construed as a potential conflict of interest.

References

1. Zhang, Q.; Bu, Y.; Li, Q. Review on fatigue problems of orthotropic steel bridge deck. *China J. Highw. Transp.* **2017**, *30*, 14–30.
2. Lyu, Z.; Jiang, X.; Qiang, X. State-of-the-art review on fatigue strengthening solutions in orthotropic steel bridge decks. *Structures* **2025**, *79*, 109597. [CrossRef]
3. Miller, T.C.; Chajes, M.J.; Mertz, D.R.; Hastings, J.N. Strengthening of a steel bridge girder using CFRP plates. *J. Bridge Eng.* **2001**, *6*, 514–522. [CrossRef]
4. Guo, D.; Wang, H.-P.; Liu, Y.-L.; Gao, W.-Y.; Dai, J.-G. Structural behavior of CFRP-strengthened steel beams at different service temperatures: Experimental study and FE modeling. *Eng. Struct.* **2023**, *293*, 116646. [CrossRef]
5. Deng, J.; Lee, M.M.K.; Moy, S.S.J. Stress Analysis of Steel Beams Reinforced with a Bonded CFRP Plate. *Compos. Struct.* **2004**, *65*, 205–215. [CrossRef]
6. Li, J.; Wang, Y.; Deng, J.; Jia, Y. Experimental study on the flexural behaviour of notched steel beams strengthened by prestressed CFRP plate with an end plate anchorage system. *Eng. Struct.* **2018**, *171*, 29–39. [CrossRef]
7. Teng, J.G.; Yu, T.; Fernando, D. Strengthening of steel structures with fiber-reinforced polymer composites. *J. Constr. Steel Res.* **2012**, *78*, 131–143. [CrossRef]
8. Yu, T.; Fernando, D.; Teng, J.G.; Zhao, X.L. Experimental study on CFRP-to-steel bonded interfaces. *Compos. Part B Eng.* **2012**, *43*, 2279–2289. [CrossRef]
9. Bai, Y.; Nguyen, T.C.; Zhao, X.L.; Al-Mahaidi, R. Environment-Assisted Degradation of the Bond between Steel and Carbon-Fiber-Reinforced Polymer. *J. Mater. Civ. Eng.* **2014**, *26*, 04014054. [CrossRef]
10. Gholami, M.; Sam, A.R.M.; Yatim, J.M.; Tahir, M.M. A review on steel/CFRP strengthening systems focusing environmental performance. *Constr. Build. Mater.* **2013**, *47*, 301–310. [CrossRef]
11. Wang, H.-T.; Bian, Z.-N.; Chen, M.-S.; Hu, L.; Wu, Q. Flexural strengthening of damaged steel beams with prestressed CFRP plates using a novel prestressing system. *Eng. Struct.* **2023**, *284*, 115953. [CrossRef]
12. Chen, T.; Gu, X.; Qi, M.; Yu, Q. Experimental study on fatigue behavior of cracked rectangular hollow-section steel beams repaired with prestressed CFRP plates. *J. Compos. Constr.* **2018**, *22*, 04018034. [CrossRef]
13. Ni, H.; Chen, Z.; Liu, Y.; Zhang, B.; Peng, H. Fatigue behavior of the bonded interface of defective steel plate reinforced with self-bonded prestressed CFRP. *Constr. Build. Mater.* **2024**, *484*, 141837. [CrossRef]
14. Deng, J.; Li, J.; Zhu, M. Fatigue behavior of notched steel beams strengthened by a prestressed CFRP plate subjected to wetting/drying cycles. *Compos. Part B Eng.* **2021**, *230*, 109491. [CrossRef]
15. Chen, Z.; Liu, Y.; He, J.; Peng, H.; Zhang, B. Prestressing effect of self-bonded prestressed CFRP for repairing steel plates with defects. *Structures* **2025**, *76*, 108994. [CrossRef]
16. Hosseini, A.; Ghafoori, E.; Al-Mahaidi, R.; Zhao, X.-L.; Motavalli, M. Strengthening of a 19th-century roadway metallic bridge using nonprestressed bonded and prestressed unbonded CFRP plates. *Constr. Build. Mater.* **2019**, *209*, 240–259. [CrossRef]
17. Hosseini, A.; Ghafoori, E.; Motavalli, M.; Nussbaumer, A.; Zhao, X.-L.; Al-Mahaidi, R.; Terrasi, G. Development of prestressed unbonded and bonded CFRP strengthening solutions for tensile metallic members. *Eng. Struct.* **2019**, *181*, 550–561. [CrossRef]
18. Li, J.; Zhu, M.; Deng, J. Flexural behaviour of notched steel beams strengthened with a prestressed CFRP plate subjected to fatigue damage and wetting/drying cycles. *Eng. Struct.* **2022**, *250*, 113430. [CrossRef]
19. Guo, D.; Zhou, H.; Wang, H.-P.; Dai, J.-G. Effect of temperature variation on the plate-end debonding of FRP-strengthened steel beams: Coupled mixed-mode cohesive zone modeling. *Eng. Fract. Mech.* **2022**, *270*, 108583. [CrossRef]
20. Guo, D.; Mao, W.-A.; Deng, J.; Zhong, M.-T.; Li, X.-D. Bond behavior of CFRP-to-steel interfaces with a typical ductile adhesive at low temperatures. *J. Constr. Steel Res.* **2025**, *231*, 109577.
21. *GB/T 2975-2018*; Steel and Steel Products-Location and Preparation of Samples and Test Pieces for Mechanical Testing. Standardization Administration of the People's Republic of China (SAC): Beijing, China, 2018.
22. *GB/T 2567-2021*; Test Methods for Properties of Resin Casting Body. China National Administration for Standardization (SAC): Beijing, China; State Administration for Market Regulation (SAMR): Beijing, China, 2021.
23. *GB/T 1447-2005*; Fiber-Reinforced Plastics Composites-Determination of Tensile Properties. China National Standardization Administration Committee: Beijing, China, 2005.

24. Li, J.; Deng, J.; Wang, Y.; Guan, J.; Zheng, H. Experimental study of notched steel beams strengthened with a CFRP plate subjected to overloading fatigue and wetting/drying cycles. *Compos. Struct.* **2019**, *209*, 634–643. [CrossRef]
25. *GB/T 8923.1-2011*; Preparation of Steel Substrates Before Application of Paints and Related Products—Visual Assessment of Surface Cleanliness—Part 1: Rust Grades and Preparation Grades of Uncoated Steel Substrates and of Steel Substrates After Overall Removal of Previous Coatings. Standardization Administration of China (SAC): Beijing, China, 2011.
26. Wang, Y.; Li, J.; Deng, J.; Li, S. Bond behaviour of CFRP/steel strap joints exposed to overloading fatigue and wetting/drying cycles. *Eng. Struct.* **2018**, *172*, 1–12. [CrossRef]
27. Ghafoori, E.; Motavalli, M. Flexural and interfacial behavior of metallic beams strengthened by prestressed bonded plates. *Compos. Struct.* **2013**, *101*, 22–34. [CrossRef]
28. Kianmofrad, F.; Ghafoori, E.; Elyasi, M.; Motavalli, M.; Rahimian, M. Rahimian. Strengthening of metallic beams with different types of pre-stressed un-bonded retrofit systems. *Compos. Struct.* **2019**, *159*, 81–95. [CrossRef]

Disclaimer/Publisher's Note: The statements, opinions and data contained in all publications are solely those of the individual author(s) and contributor(s) and not of MDPI and/or the editor(s). MDPI and/or the editor(s) disclaim responsibility for any injury to people or property resulting from any ideas, methods, instructions or products referred to in the content.

Journal of
Composites Science

MDPI

Article

Statistical Analysis of Tensile Damage of Basalt Fiber Foam Concrete Based on DBSCAN Clustering Method

Hai Yu [1,*], Changgeng Liu [1], Yangzhuang An [1], Rufeng Ma [1] and Yunpeng Liu [2]

[1] School of Civil Engineering, North Minzu University, Yinchuan 750021, China
[2] School of Materials Science and Engineering, University of Science and Technology Beijing, Beijing 100083, China
* Correspondence: yuhai1212@126.com

Abstract: The study proposes a methodology that combines digital image correlation (DIC) with cluster analysis (CA) to investigate the damage evolution and localization behavior of basalt fiber foam concrete (BFFC) under tensile loading. This method can simultaneously conduct quantitative analysis of both the process of damage accumulation and the process of damage localization. Quasi-static tensile tests were performed on specimens with different matrix densities and basalt fiber content. The full-field and full-process deformation images of the specimens were recorded by a high-resolution CCD. Cluster analysis was performed on the precise deformation data obtained from the DIC method, and damage extent factors and damage localization coefficients were defined. Statistical analysis indicates that the incorporation of basalt fibers not only effectively delays the progression of damage in foam concrete materials but also significantly enhances their initial damage threshold load and inhibits the phenomenon of damage localization in foam concrete. Compared to specimens without basalt fibers, those incorporating basalt fibers exhibited increases in the damage localization coefficients at tensile failure of 0.4, 0.33 and 0.18, respectively, under three different matrix density conditions. Therefore, the proposed DIC-CA method, in conjunction with the defined damage extent factor and damage localization coefficient, can effectively and quantitatively capture the two key dimensions of damage (accumulation extent and spatial distribution characteristics) in fiber-reinforced foam concrete under tensile loading. This provides an efficient, intuitive, quantitative analysis method for characterizing the initiation, development and localization processes of damage in similar materials.

Keywords: basalt fiber foam concrete; damage evolution; localization behavior; digital image correlation; cluster analysis

1. Introduction

As a new type of building material, foam concrete (FC) has good thermal and acoustic insulation properties due to its high porous structure [1–7]. In addition, foam concrete demonstrates excellent controllability of strength, which is achieved by precisely regulating its density. This direct correlation between strength and density enhances the flexibility of material design and ensures the best possible mechanical properties and structural stability in different application scenarios [8–11]. This enables FC to adapt flexibly to diverse environmental conditions and working demands. However, despite the many advantages of FC, its low compressive strength and significant brittleness characteristics have limited its application and widespread use [12–16].

FC is a typical brittle material that exhibits difficulty in undergoing significant plastic deformation when subjected to force. Its stress–strain curve typically experiences a rapid decline after reaching the peak, indicating low energy dissipation capacity. Domestic and international scholars have found that the study of the physical and mechanical properties of fiber-reinforced concrete materials under different environmental conditions has become a popular research topic. Fernandes et al. [17] studied the microstructure of thermally damaged concrete from real-scale reinforced concrete columns using scanning electron microscopy (SEM) and x-ray diffraction (XRD). Manica et al. [18] studied the influence of age and internal moisture on the performance of reinforced concrete walls at high temperatures. In FC, fibers can act as crack bridges and prevent further crack propagation [19]. Consequently, fiber reinforcement is frequently employed to enhance foam concrete's tensile properties [20]. It was reported that the addition of steel fibers improved the compressive and flexural strengths, and increased the ductility of lightweight concrete [21]. The addition of polypropylene (PP) fibers has been shown to increase the splitting tensile and flexural strength of foam concrete by 44% and 40%, respectively [22]. Amran et al. [23] also found that the inclusion of polypropylene fiber and silica fume enhanced the strengths of foamed concrete to levels 20–50% greater than those of the corresponding reference concrete at the specified volume of foam. Meanwhile, an increase in the tensile strength was seen at increased compressive strength. In addition to polypropylene (PP) fibers, polyvinyl alcohol (PVA) fibers also enhance the tensile behavior of foam concrete. This improvement can be attributed to their high modulus and strength, cost-effectiveness [24], and strong adhesion to the cement matrix [25]. These factors collectively confirm the advantages of incorporating PVA fibers in foam concrete applications. It has been reported that adding 0.3% (by volume) of PVA fibers can increase the splitting and bending strength of foam concrete by 27% and 76%, respectively [26]. The numerous studies mentioned above practiced the use of various supplementary cementing materials (SCM) and fibers in foam concrete. Above studies demonstrate that the incorporation of an appropriate number of fibers (such as polypropylene, steel, glass, and carbon fibers) into foam concrete can significantly enhance its mechanical properties. Fibers function as 'microtendons' during the hardening process, bridging pores and microcracks, which establishes an effective mechanism for bridging and pull-out [27]. This action delays crack propagation and improves the material's ductility and toughness. Nevertheless, studies have shown that fiber-reinforced foam concrete incorporating natural fibers, steel fibers, or polymer-based fibers is susceptible to durability issues [28], low corrosion resistance [29], and inadequate heat resistance [30]. In contrast, artificially produced ceramic fibers, such as basalt fibers, exhibit superior corrosion and temperature resistance compared to their steel, polymer-based, and natural counterparts. Osman et al. [31] found that the incorporation of basalt fibers significantly influenced the pore network of the fiber slurry, thereby enhancing its strength. Notably, the flexural strength increased by approximately 88% when the basalt fiber content was raised from 0% to 3%. The high strength and excellent thermal stability of basalt fibers have contributed to their growing application in engineering.

Although the incorporation of fibers can form a bridging link at the crack surface, which effectively delays crack propagation and improves the post-peak bearing capacity of the material, the bridging effect often exhibits spatial inhomogeneity in practical engineering applications. When fiber strength is insufficient or their distribution is non-uniform, localized fibers rupture or pull-out may occur, accelerating microcrack propagation and leading to microcrack cluster formation. FC typically comprises a substantial number of both closed-cell and open-cell structures. These pores are susceptible to becoming zones of stress concentration when subjected to external loading, which can subsequently result in the formation of microcracks. Due to the randomness of pore distribution, damage inside the structure tends to be localized rather than uniformly distributed. Additionally, the heterogeneous spatial distribution of cement paste, pores, admixtures, and fibers results in a multi-scale, multiphase discontinuous internal structure. This inherent material heterogeneity increases the susceptibility of certain regions to premature failure under load, thereby promoting damage localization. In the study of damage localization in fiber-reinforced foamed concrete, quantifying the damage and localization behavior is crucial for understanding the damage mechanism and optimizing the design. Regarding the issues, relevant literature has investigated strain localization in fiber-reinforced composites using digital Image correlation and acoustic emission techniques, a description of carbon fiber reinforced polymers (CFRP) compression damaging scenario from first micro-damages to final failure of the material and potential effect of matrix (resin) on compressive mechanical properties and the damaging scenario is given by Khalil et al. [32]. Hafiz et al. [33] focuses on the failure analysis of woven fabric carbon-reinforced polymeric composites under tensile and flexural loading. To conduct a detailed investigation acoustic emission is used to attain damage evolution under flexural loading conditions.

This paper presents an effective integration of the digital image correlation technique and cluster analysis to enable quantitative evaluation of tensile damage localization in fiber-reinforced foam concrete. DIC enables high-precision, non-contact acquisition of full-field strain data, while the clustering algorithm performs automated segmentation and damage identification within complex strain fields. This combined approach significantly enhances the accuracy of damage localization and quantification. Moreover, it accommodates the material's inherent heterogeneity and features intelligence, visualization, and a data-driven framework, thereby providing an effective tool for quantitatively tracking and intelligently assessing the damage evolution in fiber-reinforced foam concrete.

2. Materials and Methods

2.1. Raw Materials and Specimen Preparation

This study used Ningxia Yinchuan Saima ordinary portland cement, which has a 28-day compressive strength of 51 MPa. The basalt fibers were sourced from Liaoning JinShi Science and Technology Co., Ltd. (Shenyang, China), were 6 mm long, as shown in Figure 1. The primary mechanical property indices of the basalt fibers are detailed in Table 1. The blowing agent used was a commercially available animal protein blowing agent, and the mixing water was tap water from the laboratory. The specific preparation process for the test sample is as follows: (1) A foaming liquid is created by mixing the blowing agent and water in a quality ratio of 1:40. The weighed foaming liquid is injected into the foaming equipment (see Figure 2), where stable and uniform bubbles are generated by controlling the airflow rate; (2) Cement and water are mixed to form a cement slurry with a water-to-cement ratio of 1:2. The mixture is stirred for two minutes; (3) The weighed basalt fiber is uniformly mixed into the cement paste and stirred for an additional 2 min; (4) Air bubbles are introduced into the homogeneous fiber cement paste, with the foam dosage adjusted to control the density of the foam concrete. A low-speed mixing is performed for 1 min to achieve a uniform basalt fiber foam concrete slurry; (5) The basalt fiber foam concrete slurry is poured into a dumbbell-type test mold and allowed to stand at room temperature for 24 h; (6) The specimens are demolded and cured in a curing box (China Academy of Building Research, Beijing, China) for 28 days, with the temperature maintained at 20 °C and the relative humidity at 95%.

Table 1. Main properties of basalt fiber.

Diameter/μm	Tensile Strength/MPa	Elastic Modulus/GPa	Elongation/%	Density/(kg·m^{-3})
9–17	3000–4800	85–110	3.0–3.5%	2.6–2.8

Figure 1. Basalt fiber physical picture.

Figure 2. Foaming machine physical picture.

2.2. Design of Tensile Specimens

Three groups of uniaxial quasi-static tensile specimens with different matrix densities were designed for testing. The tensile specimens were dumbbell-shaped (see Figure 3 for detailed dimensions), with each group containing six specimens with different fiber content but the same matrix density. This resulted in a total of 18 quasi-static tensile specimens. Two test variables were considered: foam concrete density and basalt fiber volume dosing. The designed densities of foam concrete were 900 kg/m^3, 1050 kg/m^3, and 1200 kg/m^3, while the basalt fiber volume dosages were 0%, 0.1%, 0.2%, 0.3%, 0.4%, and 0.5%. The length of the basalt fiber was 6 mm, and the physical drawings of a set of specimens are provided in Figure 4.

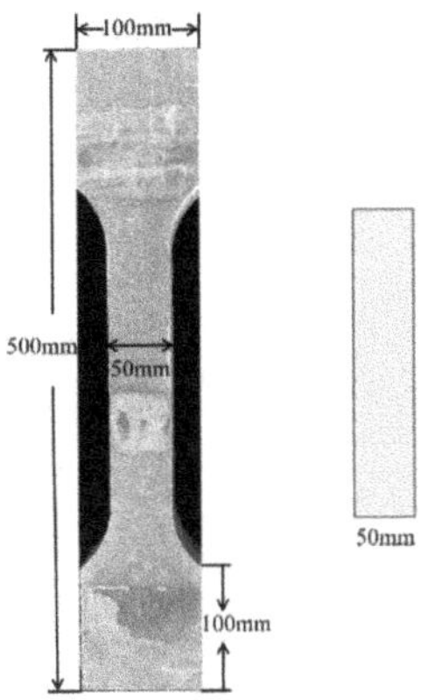

Figure 3. Dimensions of quasi-static tensile specimens.

Figure 4. Physical drawing of quasi-static tensile specimen and speckle.

2.3. Uniaxial Quasi-Static Tensile Test

The quasi-static tensile test was performed using an MTS Exceed E44.304 microcomputer-controlled electronic universal materials testing machine (MTS, Eden Prairie, MN, USA) (see Figure 3). Due to the brittle nature of foam concrete, the experiment used displacement loading control and applied a relatively low loading rate of 0.2 mm/min. This study integrates the digital image correlation method with cluster analysis to quantitatively assess tensile damage localization in fiber-reinforced foam concrete. To facilitate this, a CCD camera (China Daheng (Group) Co., Ltd., Beijing, China) was positioned in front of the tensile specimen to capture real-time surface images under varying loading conditions. The CCD image acquisition device has a resolution of 3088 × 2064 pixels, with an imaging area of 420 mm × 280 mm. When analyzing the experimental results, the size of the speckles on the specimen surface has a significant influence on the measurement accuracy of the DIC method. The size and uniformity of the speckles produced by different fabrication techniques can vary considerably. According to references [34–36], the ideal speckle size is between 4 and 6 pixels. Considering the specimen size in this experiment, we used a method of randomly applying spots with a black marker to create speckles, thereby ensuring randomness and uniformity in their distribution (see Figure 5). In the DIC calculation process of this article, the size of the displacement calculation subset is 31 × 31, and the calculation step size is 10.

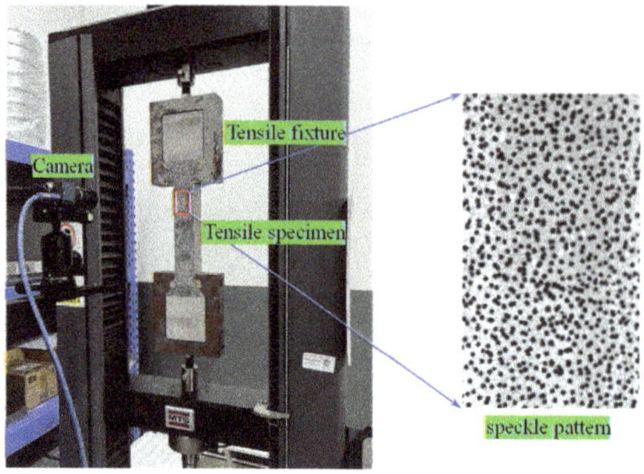

Figure 5. Schematic diagram of the testing system and speckle pattern of the specimen surface.

2.4. DBSCAN Clustering Method

DBSCAN (Density-Based Spatial Clustering of Applications with Noise) is a very popular and powerful unsupervised machine learning algorithm that is primarily used to discover clusters in data [37]. Unlike other clustering algorithms such as K-means, DBSCAN does not require the number of clusters to be pre-specified and is capable of recognizing clusters of arbitrary shapes, as well as marking data points that are considered noisy. The core concept of DBSCAN is to define clusters based on density. The core principle relies on two important parameters: (1) Epsilon (ε): defines the distance threshold from a given point to another point. If the distance between two points is less than or equal to ε, they are considered "neighboring". (2) MinPts (λ): Indicates the minimum number of neighboring points (including the point itself) required for a point to be considered as a core point. For more details, refer to reference [38,39]. An example plot of the DBSCAN clustering method is given in Figure 6.

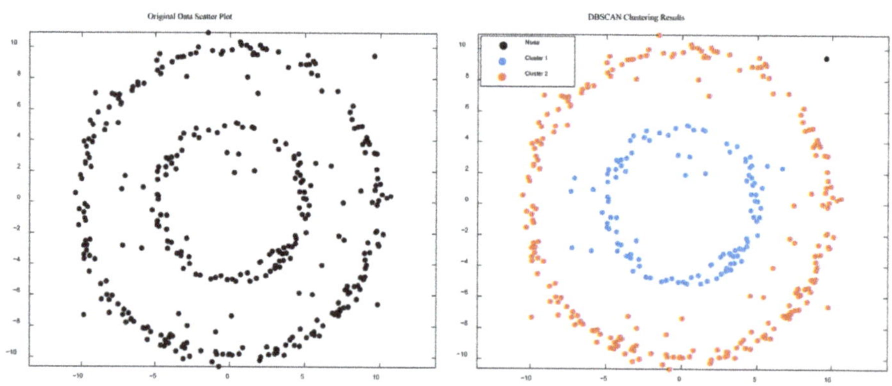

Figure 6. DBSCAN clustering example graph (**left**: original data scatter plot, **right**: DBSCAN clustering results).

In this study, a substantial amount of strain data acquired through the digital image correlation method will be systematically analyzed utilizing the cluster analysis technique. By performing cluster analysis on the strain field, key feature regions in the damage evolution process will be identified, facilitating the quantitative identification and assessment of damage localization behavior in basalt fiber foam concrete under uniaxial tensile loading. Additionally, this research aims to investigate the effect of basalt fiber incorporation on damage localization behavior. By comparing the clustering results of specimens with varying fiber volume fractions, the role of basalt fibers in delaying damage progression, dispersing strain concentration, and enhancing overall structural crack resistance will be analyzed.

3. Results and Discussion

3.1. Tensile Damage Analysis

The tests were performed using the DIC method to obtain full-time and full-field displacement and strain information from the specimen surface during the loading process. Under tensile loading, the tensile strain in the vertical direction plays a decisive role in the tensile damage process. Therefore, vertical strain (ε_{yy}) was selected as the analysis factor, and seven characteristic points (A–G, Figure 7) were chosen from the displacement-load curve to analyze the tensile damage process of BFFC. It is important to note that the size of the strain calculation window significantly influences the results of strain field calculations using DIC. This study reflects the localization of damage (strain localization) in the material under tensile loading by accurately calculating strains. Figure 8 illustrates the strain results obtained from different strain windows along the yellow dashed line on the specimen surface. It demonstrates that a smaller strain window is more sensitive to significant strain regions and can accurately reflect their deformation. In contrast, a larger strain window tends to over-smooth the displacement data, leading to distorted strain information. Consequently, this paper uses a strain calculation window size of 5×5. Figure 9 shows the distribution of strain evolution in the vertical direction at each characteristic point (A–G, with loads gradually increasing) of the BFFC-900 specimen without basalt fiber. Comparative analyses indicate that the tensile damage process of the BFFC can be categorized into three distinct stages: the microcrack initiation stage, the microcrack synergy stage, and the main crack development stage. Microcrack initiation stage (B–C): The overall vertical tensile strain level of the specimen is low at this stage. However, a more obvious strain concentration area has appeared in the local area at the edge of the specimen, indicating that microcracks have begun to emerge there. Microcrack synergistic stage (C–E): In this stage, with the increase in load, multiple microcracks form and develop synergistically in the specimen. Significant strain concentration occurs in the local area (especially in the main crack formation area), which is much higher than in other areas. In the latter part of this stage, main cracks gradually form. Main crack development stage (E–G): The main crack forms in the middle of the specimen near the left edge. The vertical tensile strain at this position is significantly higher than that in other areas. With the continuous load increase, this strain difference expands dramatically, which ultimately leads to specimen fracture at the main crack.

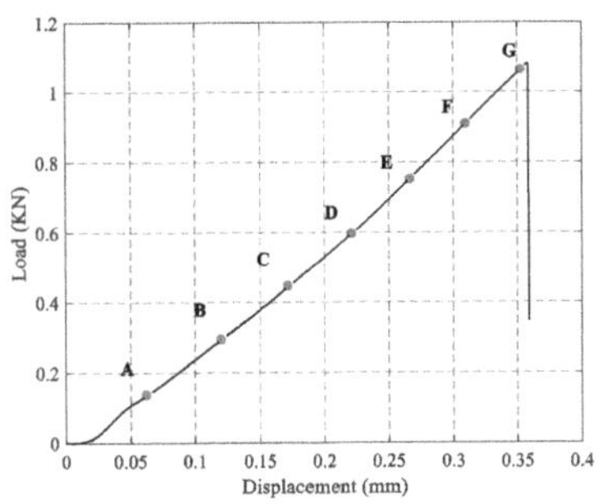

Figure 7. The displacement-load diagram of the BFFC-900 specimen without basalt fiber.

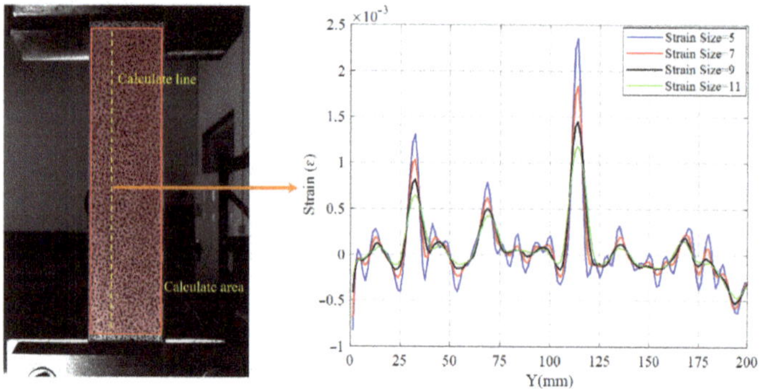

Figure 8. The strain results calculated under different strain windows along the yellow dotted line on the specimen surface.

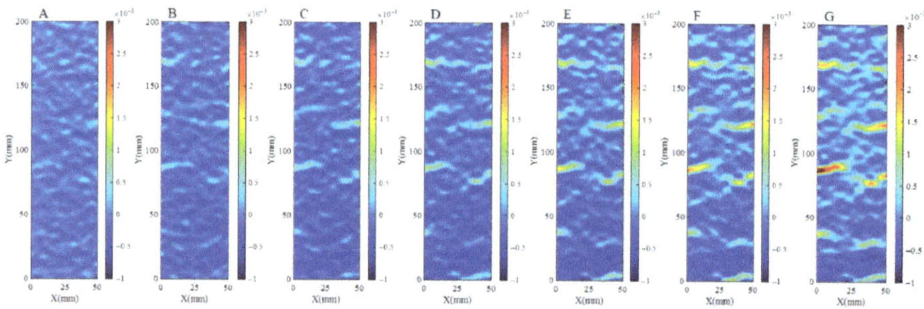

Figure 9. The vertical strain evolution at each characteristic point ((**A–G**) load increasing gradually) of BFFC-900 specimen without basalt fiber reinforcement.

3.2. *Ultimate Tensile Strength*

The ultimate tensile strength of all specimens was determined by performing quasi-static tensile tests. Figure 10 shows the ultimate tensile strength of BFFCs and the percentage increase in ultimate tensile strength compared to specimens without basalt fiber reinforcement. The results demonstrate that the addition of basalt fibers significantly enhances the ultimate tensile strength of BFFCs with different matrix densities. However, the enhancement patterns for BFFCs with different matrix densities are not entirely consistent. For BFFC-900 and BFFC-1050, which have relatively low matrix densities, the tensile strength does not increase with the addition of basalt fibers. For BFFC-900, the tensile strength reaches its maximum value at a basalt fiber content of 0.2%, reflecting a 29.57% increase compared to the ultimate tensile strength of foam concrete without basalt fiber. Conversely, when the basalt fiber content increases to 0.3%, the tensile strength rises by only 11.27%. Similarly, the tensile strength of BFFC-1050 also peaks at a basalt fiber content of 0.2%, showing a 71.89% increase compared to foam concrete without basalt fiber. However, at a basalt fiber content of 0.3%, the increase in tensile strength diminishes to 42.48%. The above phenomena can be analyzed from two perspectives. Firstly, it is challenging to control the density of foam concrete precisely during the sample preparation process, leading to fluctuations in the actual density of the prepared foam concrete relative to the designed density within a certain range. Since the strength of foam concrete is closely related to its density, this discrepancy can significantly impact the material's performance. Secondly, for foam concrete with a relatively low density and high porosity, fibers cannot bond effectively with the matrix. This means that the bridging effect of fibers is not fully utilized, thus diminishing the correlation between strength and fiber content. Conversely, for BFFC-1200, which has a relatively high matrix density, there is a positive correlation between ultimate tensile strength and fiber content. As the fiber content increases, tensile strength also increases, reaching a maximum value at 0.4% fiber content, which is 61.57% higher than that of foam concrete without basalt fiber. However, when the basalt fiber content reaches 0.5%, the increase in tensile strength diminishes. These results suggest that the optimal basalt fiber content is not fixed, but depends on the matrix density of the foam concrete. In the experiments conducted in this study, the optimal basalt fiber content for the three matrix densities ranged from 0.2% to 0.4%.

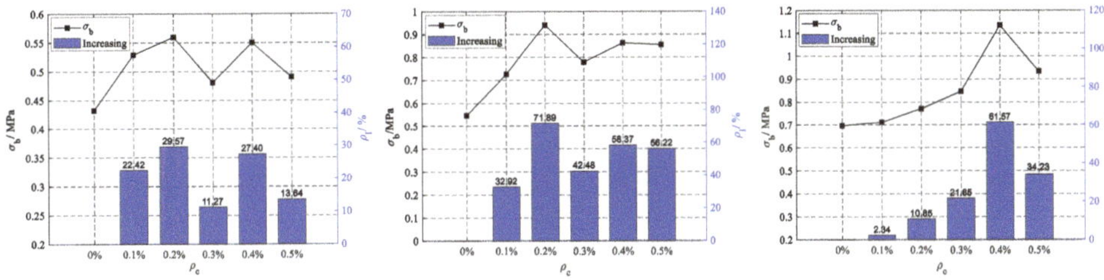

(a) (b) (c)

Figure 10. Variation in ultimate tensile strength. (**a**) BFFC-900, (**b**) BFFC-1050, (**c**) BFFC-1200.

As illustrated in Figure 11, the optical images of the fracture surfaces of all the specimens after fracturing are presented. It has been observed that, as the content of basalt fibers increases, the corresponding fracture surfaces become increasingly irregular. This phenomenon can be attributed to the fact that the process of fiber pull-out requires the expenditure of additional energy to overcome interfacial friction and adhesion forces. During this process, the fibers are pulled out of the matrix, leaving pits (where the fibers were pulled out) and protrusions (fiber tips remaining in the matrix) on the fracture surface. The numerous fiber pull-out marks are the primary cause of the rough and uneven fracture surface. Additionally, due to factors such as the strength distribution of the fibers themselves, differences in their orientation within the matrix, varying local stress states, and the presence of micro-defects, the fracture location for different fibers does not typically occur on the same plane. This significantly increases the irregularity of the fracture. Figure 12 shows a magnified image of the fracture surface of three samples with different matrix densities without basalt fiber reinforcement. Through appropriate image processing, it can be seen that, under the same preparation process, the main difference between foam concrete with different matrix densities is the size of the foam particles. The higher the matrix density, the smaller the corresponding foam particle size.

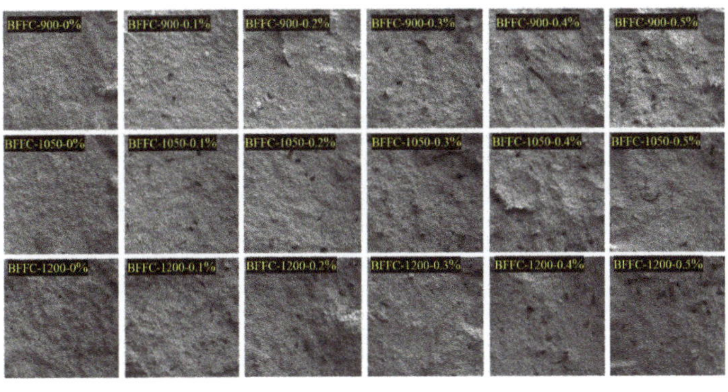

Figure 11. Optical image of the fracture surface of all specimens.

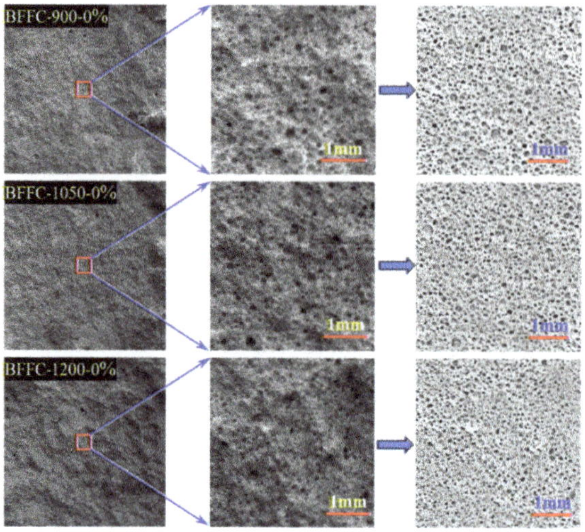

Figure 12. Local magnified images of fracture surfaces of specimens with different matrix densities without basalt fiber addition.

3.3. Statistical Analysis of the Strain Field Based on DBSCAN

To quantify the damage accumulation and damage localization of basalt fiber foam concrete under uniaxial tensile loading, cluster analysis was performed on a large amount of strain data obtained from digital image correlation (DIC) using the DBSCAN method. Specifically, for each load level shown in Figure 9, the top 181 strain points (3% of all calculated points) in the vertical direction were subjected to DBSCAN clustering analysis. The evolution of the top 181 strain points effectively reflects the strain localization process in the specimen. The distribution of these vertical tensile strain points also reflects the damage evolution process of the specimen [40,41]. Therefore, statistical analysis of the strain of the top 181 points was conducted to quantify the damage evolution process of the specimen as the load increases, as well as the damage extent and localization extent of the specimen at different load levels. For this purpose, two indicators (the damage extent factor and the damage localization factor) were defined to characterize the damage of the specimen:

The damage extent factors (D_f) is used to characterize the increment of vertical tensile strain in BFFCs, and is defined as follows:

$$D_f = \bar{\varepsilon}_i / \bar{\varepsilon}_{\max}$$

$$\bar{\varepsilon}_i = \frac{1}{M}\sum_{j=1}^{M} (\varepsilon_{yy})_j - \frac{1}{N}\sum_{j=1}^{N} (\varepsilon_{yy})_j.$$

where $\bar{\varepsilon}_i$ is the difference between the average vertical strain of the top M (In this article, $M = N$*0.03) points and the average vertical strain of all the points in the analysis area; N is the number of points in the analysis area; ε_{yy} is the vertical strain of point j; $\bar{\varepsilon}_{\max}$ is the maximum of $\bar{\varepsilon}_i$.

The damage localization factor (L_f) is used to characterize the centralization of the distribution of BFFCs in the vertical direction, and is defined as follows:

$$L_f = s_x / s_{x\max}$$

$$s_x = \sum_{n=1}^{n_x} \sum_{j=1}^{k_n} |y_{nj} - y_{nc}|$$

where y_{nj} is the y-coordinate in the nth cluster of the top M points; y_{nc} is the y-coordinate of cluster center in the nth cluster; k_n is the number of points contained in the top M points in the nth cluster; n_x is the number of clusters; $s_{x\max}$ is the maximum of s_x.

Section 2.4 of this paper states that the core principle of DBSCAN relies on two important parameters: Epsilon (ε) and MinPts (λ). According to the strain data characteristics in this paper, take ε = 10, λ = 3. Figure 13 shows the DBSCAN clustering results for the top 181 tensile strain points in the vertical direction at each load level corresponding to Figure 9. In each figure, different colored regions represent different strain classes and red crosses indicate the cluster centers corresponding to the same color class. Under the initial loading condition, it can be seen that the high tensile strain points on the specimen surface exhibit a discrete distribution, indicating that no obvious damage regions have yet formed at this stage. As the load increases (stages B to E), the high tensile strain points begin to aggregate and evolve in localized regions, clearly showing the emergence and continuous development of multi-regional damage. In the later loading stage (stages E to G), the main damage zone gradually forms and dominates the damage evolution process, while the development of other secondary damage zones tends to slow down or stagnate. Ultimately, the specimen fractures and fails in the main damage zone.

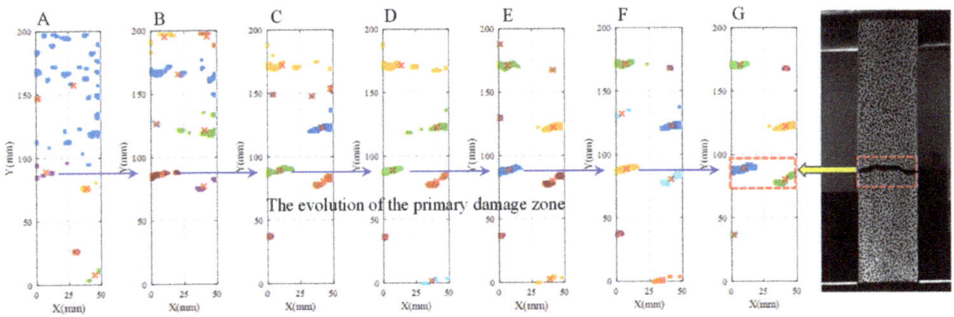

Figure 13. The DBSCAN clustering results of the vertical stretching strain points under each load level corresponding to Figure 9.

Figure 14 shows the load-dependent curves of D_f and L_f under different matrix densities and basalt fiber content. The results indicate that, despite differences in matrix density and fiber content, the load-dependent curves of D_f and L_f in all specimens exhibit similar trends. For D_f: it can be divided into two stages. Stage I (low-value fluctuation stage): D_f remains at a relatively low level overall, fluctuating within a certain numerical range. This indicates that the specimen has not yet undergone significant macroscopic damage at this load stage. Stage II (rapid growth stage): D_f significantly increases and exhibits a rapid growth trend with increasing load. This indicates that the specimen has entered a distinct stage of damage development, with damage continuing to accumulate and intensify. For L_f: it is also divided into two stages. Stage I (rapid decline stage): L_f decreases significantly with increasing load. This change characteristic indicates that strain localization phenomena have occurred in the specimen during this stage, and the strain localization regions exhibit a multi-point concurrent trend. Stage II (slow decline stage): L_f has decreased to a relatively low level, with a significantly slowed decline rate. This marks the significant development of damage localization, primarily due to the formation and continued development of the dominant damage zone, while the development of secondary damage zones has significantly slowed or even stagnated.

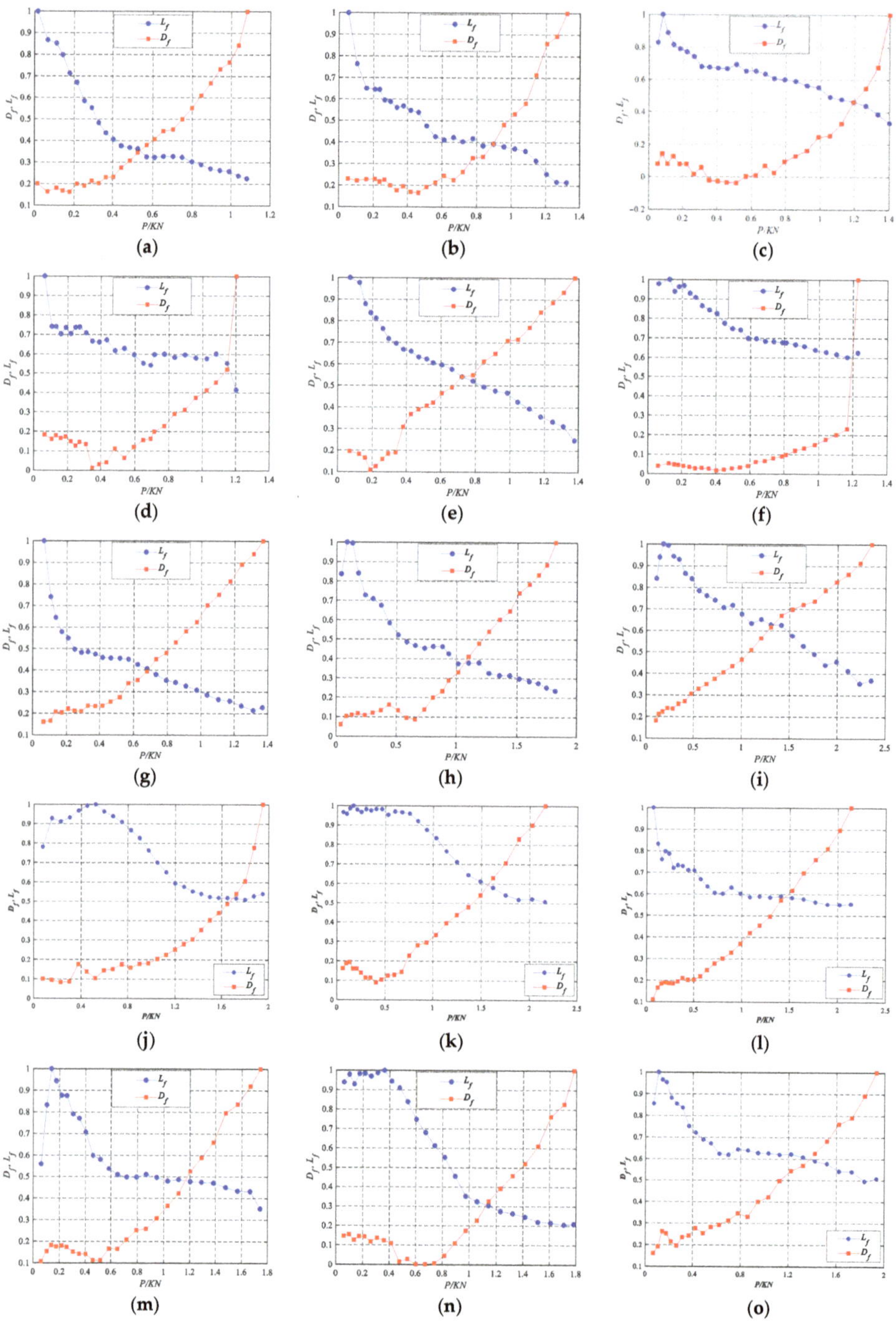

Figure 14. *Cont.*

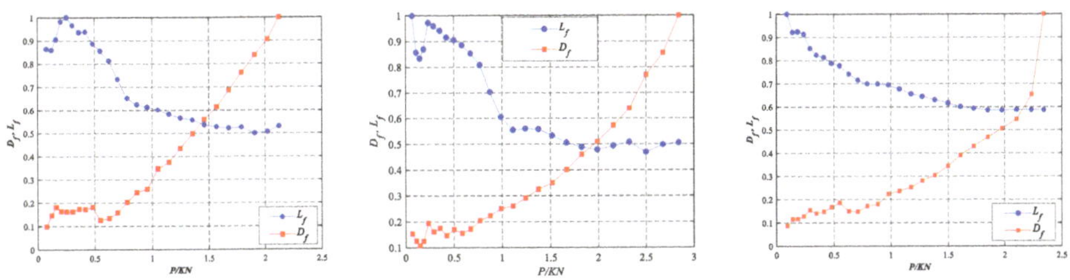

(p) (q) (r)

Figure 14. Variation of D_f and L_f with increment of load for all specimens. (**a**) BFFC-900-0%, (**b**) BFFC-900-0.1%, (**c**) BFFC-900-0.2%, (**d**) BFFC-900-0.3%, (**e**) BFFC-900-0.4%, (**f**) BFFC-900-0.5%, (**g**) BFFC-1050-0%, (**h**) BFFC-1050-0.1%, (**i**) BFFC-1050-0.2%, (**j**) BFFC-1050-0.3%, (**k**) BFFC-1050-0.4%, (**l**) BFFC-1050-0.5%, (**m**) BFFC-1200-0%, (**n**) BFFC-1200-0.1%, (**o**) BFFC-1200-0.2%, (**p**) BFFC-1200-0.3%, (**q**) BFFC-1200-0.4%, (**r**) BFFC-1200-0.5%.

3.3.1. Effect of Basalt Fibers on the D_f

Figure 15 shows a comparison of the damage-load curves of specimens with varying basalt fiber content. The damage development process (damage extent-load curve) of basalt fiber foam concrete with different matrix densities can be divided into two stages. In stage I: the damage extent-load curves of specimens with different basalt fiber content all exhibit a gradual upward trend and have similar shapes. During this stage, the damage extent of basalt fiber foam concrete with different matrix densities are all below 0.4, indicating a low level and suggesting that the fibers have a very limited inhibitory effect on material damage evolution. Stage II: As the load increases, the rate of increase in the damage extent-load curves of specimens with different basalt fiber content accelerates significantly. At this point, the differentiated effects of fiber content on damage control become evident. The same basic pattern is observed under three different matrix densities: at the same load level, the higher the fiber content, the smaller the corresponding damage variable, indicating a lower degree of material damage. Conversely, a decrease in fiber content results in an increase in the damage variable and a faster growth rate of the damage. Consequently, the incorporation of basalt fibers not only significantly delays the material's damage progression but also effectively increases its initial damage threshold load.

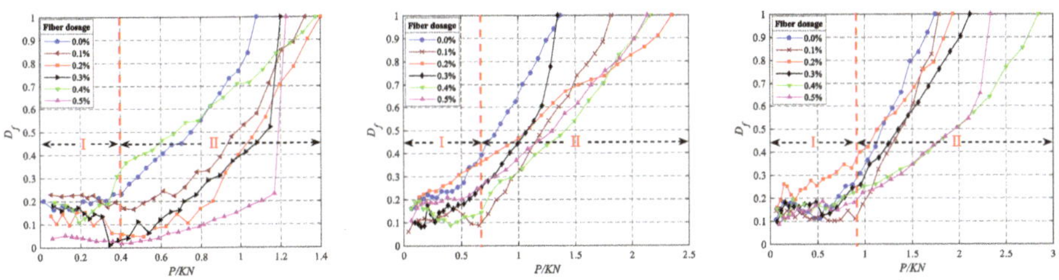

(a) (b) (c)

Figure 15. Comparison chart for variation of D_f with increment of load. (**a**) BFFC-900, (**b**) BFFC-1050, (**c**) BFFC-1200.

3.3.2. Effect of Basalt Fibers on the L_f

Figure 16 shows the variation curve of the localization coefficient with load. Similar to the pattern described in Section 3.3.1, this curve can also be divided into two distinct stages: Stage I: During this stage, the discretely distributed high tensile strain points gradually converge toward the local region of the specimen, forming a potential damage concentration zone. In this stage, the incorporation of fibers effectively suppresses the localization of strain. At the same load level, the localization coefficient increases with the increase in fiber content, indicating that fibers disperse stress and promote a more uniform stress distribution. Stage II: After entering this stage, the rate of decrease in the localization coefficient with increasing load significantly slows down. The main reason is that the primary damage zone (the primary crack zone) has essentially formed in this stage. Further increases in load primarily lead to the continued expansion of the main damage zone, while other secondary damage zones stabilize as stress is released. It is worth noting that fibers also have a certain inhibitory effect on the development of damage in the main damage zone during this stage. Figure 16 also shows that, at this stage, the higher the fiber content, the larger the corresponding localization coefficient, reflecting that higher fiber content can more effectively inhibit the rapid expansion of the main damage zone. Additionally, Figure 17 presents bar charts showing how the localization coefficient varies with fiber content for three different matrix density specimens at the moment of failure. This figure visually compares the influence of fiber content on damage localization at the moment of failure under different matrix densities.

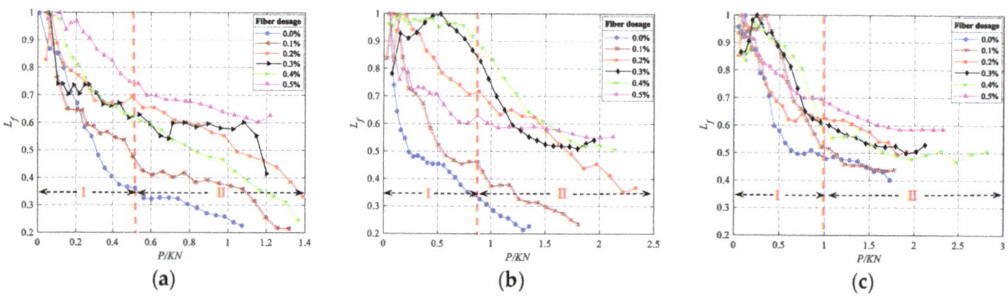

Figure 16. Comparison chart for variation of L_f with increment of load. (**a**) BFFC-900, (**b**) BFFC-1050, (**c**) BFFC-1200.

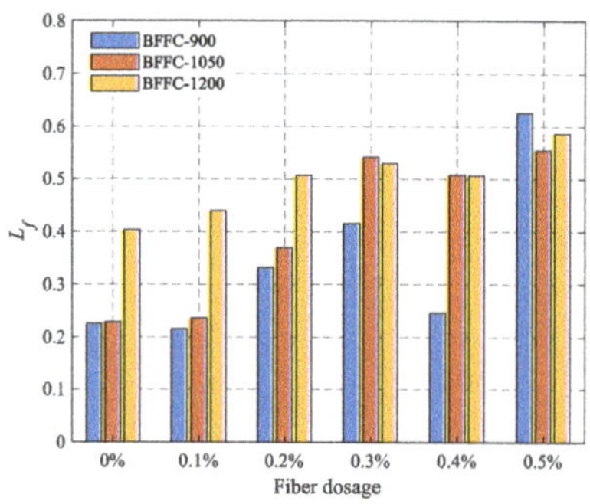

Figure 17. Bar chart of L_f at tensile failure.

4. Conclusions

This study innovatively combines the DIC method with CA to propose a damage localization quantification analysis method named DIC-CA. This method is applied to systematically investigate the damage evolution process of basalt fiber foam concrete under quasi-static tensile loading conditions. Based on the clustering analysis results of the DIC full-field deformation data, two key damage characterization parameters were defined: the damage degree factor and the damage localization coefficient. These parameters are used to quantify the overall damage accumulation state of the material and the non-uniform spatial concentration of damage, respectively. By analyzing the evolution patterns of these parameters during the tensile process, the following main conclusions were drawn:

(1) The experimental results of this study indicate that the addition of basalt fiber significantly enhances the ultimate tensile strength of basalt fiber foam concrete. The experiments demonstrate that, under three different matrix densities, the maximum tensile strength of specimens containing basalt fiber increased by 29.57%, 71.89%, and 61.57%, respectively, compared to the control group without fiber addition. It is worth noting that the experimental conclusions indicate that the optimal addition ratio of basalt fibers is not a fixed value but is closely related to the matrix density of the foam concrete. Under the three matrix density conditions studied in this research, the optimal addition ratio range of basalt fibers is 0.2–0.4%.

(2) The statistical analysis results of this experiment show that the addition of basalt fibers not only effectively delays the damage process of the material but also increases the initial damage threshold load of the material. At the same time, the fibers disperse stress through their bridging action, promoting a more uniform stress distribution and effectively inhibiting the localization and concentration of damage (especially the rapid development of the main damage zone).

(3) The damage degree factor and damage localization coefficient defined by the DIC-CA method in this paper can synchronously and quantitatively characterize two key dimensions of material damage: the former objectively reflects the overall damage accumulation of the material, while the latter precisely quantifies the degree of non-uniform spatial concentration of damage.

(4) Although the DIC-CA method proposed in this paper demonstrates certain advantages in the quantitative analysis of material damage extent and strain localization, it also exhibits certain limitations. These limitations lie in the statistical analysis results being relatively sensitive to the two parameters of the DBSCAN clustering method, and to some extent also being influenced by the DIC computational parameters. The optimization of parameter combinations requires further research and exploration.

Author Contributions: Conceptualization, H.Y.; methodology, H.Y.; software, C.L. and Y.A., validation, C.L. and Y.L.; formal analysis, H.Y.; resources, R.M.; data curation, R.M.; writing—original draft preparation, Y.A.; writing—review and editing, H.Y.; visualization; Y.A. and Y.L.; supervision, Y.A.; funding acquisition, H.Y. All authors have read and agreed to the published version of the manuscript.

Funding: The authors would like to thank the financial support from the Natural Science Foundation of Ningxia (Project No. 2026A1459) and the Young and Middle-aged Key Personnel Project of North Minzu University (Project No. 2025BG265).

Institutional Review Board Statement: Not applicable.

Data Availability Statement: The original contributions presented in this study are included in the article. Further inquiries can be directed to the corresponding author.

Conflicts of Interest: The authors declare no conflicts of interest.

Abbreviations

The following abbreviations are used in this manuscript:

BFFC	Basalt Fiber Foam Concrete
FC	Foam Concrete
DIC	Digital Image Correlation
CA	Cluster Analysis

References

1. Hou, L.; Li, J.; Lu, Z. Influence of foaming agent on cement and foam concrete. *Constr. Build. Mater.* **2021**, *280*, 122399. [CrossRef]
2. Yao, T.; Tian, Q.; Zhang, M. Experimental research on the preparation and properties of foamed concrete using recycled waste concrete powder. *Constr. Build. Mater.* **2023**, *407*, 133370. [CrossRef]
3. Dhasindrakrishna, K.; Ramakrishnan, S.; Pasupathy, K. Collapse of fresh foam concrete: Mechanisms and influencing parameters. *Cem. Concr. Comp.* **2021**, *5*, 104151. [CrossRef]
4. Xiong, Y.; Zhang, C.; Chen, C. Effect of superabsorbent polymer on the foam-stability of foamed concrete. *Cem. Concr. Comp.* **2022**, *127*, 104398.
5. Majeed, S.; Mydin, M.; Bahrami, A. Development of ultra-lightweight foamed concrete modified with silicon dioxide (SiO_2) nanoparticles: Appraisal of transport, mechanical, thermal, and microstructural properties. *J. Mater. Res. Technol.* **2024**, *30*, 3308–3327. [CrossRef]
6. Mydin, M.; Abdullah, M.; Sor, N. Thermal conductivity, microstructure and hardened characteristics of foamed concrete composite reinforced with raffia fiber. *J. Mater. Res. Technol.* **2023**, *26*, 15. [CrossRef]
7. Mydin, M.; Jagadesh, P.; Bahrami, A. Use of calcium carbonate nanoparticles in production of nano-engineered foamed concrete. *J. Mater. Res. Technol.* **2023**, *26*, 18. [CrossRef]
8. Liu, Y.; Wang, L.; Cao, K.; Sun, L. Review on the durability of polypropylene fibre-reinforced concrete. *Adv. Civ. Eng.* **2021**, *2021*, 6652077. [CrossRef]
9. Pakravan, H.; Latifi, M.; Jamshidi, M. Hybrid short fiber reinforcement system in concrete: A Review. *Constr. Build. Mater.* **2017**, *142*, 280e94. [CrossRef]
10. Tambichik, M.; Abdul Samad, A.; Mohamad, N.; Mohd Ali, A.; Othuman Mydin, M.; Mohd Bosro, M. Effect of combining palm oil fuel ash (POFA) and rice husk ash (RHA) as partial cement replacement to the compressive strength of concrete. *Int. J. Integr. Eng.* **2018**, *10*, 61–67. [CrossRef]
11. Awang, H.; Mydin, M.; Roslan, A. Effects of fibre on drying shrinkage, compressive and flexural strength of lightweight foamed concrete. *Adv. Mater. Res.* **2012**, *587*, 144e9. [CrossRef]
12. Celik, A.; Tunc, U.; Bahrami, A. Use of waste glass powder toward more sustainable geopolymer concrete. *J. Mater. Res. Technol.* **2023**, *24*, 8533–8546. [CrossRef]
13. Xiong, Y.; Li, B.; Chen, C. Properties of foamed concrete with $Ca(OH)_2$ as foam stabilizer. *Cem. Concr. Comp.* **2021**, *118*, 103985.
14. Yao, T.; Tian, Q.; Zhang, M. Laboratory investigation of foamed concrete prepared by recycled waste concrete powder and ground granulated blast furnace slag. *J. Clean. Prod.* **2023**, *426*, 139095. [CrossRef]
15. Shang, X.; Qu, N.; Li, J. Development and functional characteristics of novel foam concrete. *Constr. Build. Mater.* **2022**, *324*, 126666. [CrossRef]
16. Falliano, D.; Restuccia, L.; Gugliandolo, E. A simple optimized foam generator and a study on peculiar aspects concerning foams and foamed concrete. *Constr. Build. Mater.* **2020**, *268*, 121101. [CrossRef]
17. Fernandes, B.; Gil, A.M.; Bolina, F.L.; Tutikian, B.F. Thermal damage evaluation of full scale concrete columns exposed to high temperatures using scanning electron microscopy and X-ray diffraction. *DYNA* **2018**, *85*, 123–128. [CrossRef]
18. Manica, G.C.; Bolina, F.L.; Tutikian, B.F.; Oliveira, M.; Moreir, M.A. Influence of curing time on the fire performance of solid reinforced concrete plates. *J. Mater. Res. Technol.* **2020**, *9*, 2506–2512. [CrossRef]
19. Li, Y.; Gu, Z.; Zhao, B.; Zhang, J.; Zou, X. Experimental Study on Mechanical Properties of Basalt Fiber Concrete after Cryogenic Freeze-Thaw Cycles. *Polymers* **2023**, *15*, 196. [CrossRef]
20. Grzymski, F.; Musiał, M.; Trapko, T. Mechanical properties of fiber reinforced concrete with recycled fibers. *Constr. Build. Mater.* **2019**, *198*, 323–331. [CrossRef]
21. Mohamed, A.; Tayeh, B.; Aisheh Abu, Y.; Salih, M. Exploring the performance of steel fiber reinforced lightweight concrete: A case study review. *Case Stud. Constr. Mater.* **2023**, *18*, e01968. [CrossRef]
22. Jaini, Z.; Rum, R.; Boon, K. Strength and fracture energy of foamed concrete incorporating rice husk ash and polypropylene mega-mesh 55. *IOP Conf. Ser. Mater. Sci. Eng.* **2017**, *248*, 012005. [CrossRef]
23. Mugahed Amran, Y.; Alyousef, R.; Alabduljabbar, H.; Khudhair, M.; Hejazi, F.; Alaskar, A.; Alrshoudi, F.; Siddika, A. Performance properties of structural fibred foamed concrete. *Results Eng.* **2020**, *5*, 100092. [CrossRef]
24. Sun, M.; Zhu, J.; Sun, T.; Chen, Y.; Li, X.; Yin, W.; Han, J. Multiple effects of nanoCaCO3 and modified polyvinyl alcohol fiber on flexure-tension-resistant performance of engineered cementitious composites. *Constr. Build. Mater.* **2021**, *303*, 124426. [CrossRef]
25. Pakravan, H.; Ozbakkaloglu, T. Synthetic fibers for cementitious composites: A critical and in-depth review of recent advances. *Constr. Build. Mater.* **2019**, *207*, 491–518. [CrossRef]
26. Raj, B.; Sathyan, D.; Madhavan, M.; Raj, A. Mechanical and durability properties of hybrid fiber reinforced foam concrete. *Constr. Build. Mater.* **2020**, *245*, 118373. [CrossRef]
27. Blythe, A.; Fox, B.; Nikzad, M.; Eisenbart, B.; Chai, B.X.; Blanchard, P. Evaluation of the failure mechanism in polyamide nanofiber veil toughened hybrid carbon/glass fiber composites. *Materials* **2022**, *15*, 8877. [CrossRef]

28. Ahmad, F.; Choi, H.; Park, M. A review: Natural fiber composites selection in view of mechanical, light weight, and economic properties. *Macromol. Mater. Eng.* **2015**, *300*, 10–24. [CrossRef]
29. Frazao, C.; Barros, J.; Camoes, A.; Alves, A.; Rocha, L. Corrosion effects on pullout behavior of hooked steel fibers in self-compacting concrete. *Cem. Concr. Res.* **2016**, *79*, 112–122. [CrossRef]
30. Shihada, S. Effect of polypropylene fibers on concrete fire resistance. *J. Civ. Eng. Manag.* **2021**, *17*, 259–264. [CrossRef]
31. Li, J.; Yu, Y.; Kim, T. Unveiling the underlying mechanisms of tensile behavior enhancement in fiber reinforced foam concrete. *Constr. Build. Mater.* **2023**, *398*, 132509. [CrossRef]
32. Hamdi, K.; Moreau, G.; Aboura, Z. Digital image correlation, acoustic emission and in-situ microscopy in order to understand composite compression damage behavior. *Compos. Struct.* **2020**, *258*, 113424. [CrossRef]
33. Ali, H.Q.; Tabrizi, I.E.; Khan, R.M.A.; Tufani, A.; Yildiz, M. Microscopic analysis of failure in woven carbon fabric laminates coupled with digital image correlation and acoustic emission. *Compos. Struct.* **2019**, *230*, 111515. [CrossRef]
34. Zhou, P. Subpixel displacement and deformation gradient measurement using digital image/speckle correlation (DISC). *Opt. Eng.* **2001**, *40*, 1613–1620. [CrossRef]
35. Yu, H.; Guo, R.; Xia, H. Application of the mean intensity of the second derivative in evaluating the speckle patterns in digital image correlation. *Opt. Laser Eng.* **2014**, *60*, 32–37. [CrossRef]
36. Pan, B.; Lu, Z.; Xie, H. Mean intensity gradient: An effective global parameter for quality assessment of the speckle patterns used in digital image correlation. *Opt. Laser Eng.* **2010**, *48*, 469–477. [CrossRef]
37. Ester, M. A density-based algorithm for discovering clusters in large spatial databases with noise. *Proc. Int. Conf. Knowl. Discov. Data Min.* **1996**, *96*, 226–231.
38. Luchi, D.; Rodrigues, A.L.; Varejão, F.M. Sampling approaches for applying DBSCAN to large datasets. *Pattern Recogn. Lett.* **2019**, *117*, 90–96. [CrossRef]
39. Hanif, M.; Seo, S.; Tran, H.; Khol, S. A novel method for acoustic emission source location in CFRP-concrete debonding using ΔT mapping and DBSCAN algorithm. *Measurement* **2024**, *236*, 115097. [CrossRef]
40. Zhao, Y.; Wang, L.; Lei, Z. Study on bending damage and failure of basalt fiber reinforced concrete under freeze-thaw cycles. *Constr. Build. Mater.* **2018**, *163*, 460–470. [CrossRef]
41. Zhang, H.; Huang, G.; Song, H. Experimental characterization of strain localization in rock. *Geophys. J. Int.* **2013**, *194*, 1554–1558. [CrossRef]

Disclaimer/Publisher's Note: The statements, opinions and data contained in all publications are solely those of the individual author(s) and contributor(s) and not of MDPI and/or the editor(s). MDPI and/or the editor(s) disclaim responsibility for any injury to people or property resulting from any ideas, methods, instructions or products referred to in the content.

Journal of
Composites Science

MDPI

Article

Degradation of Tensile Properties in CFRCM Composites Under Anodic Polarization: Role of Standardized Electrolyte Solutions

Miaochang Zhu [1], Yawen Zhang [1], Haorui Chen [1], Jun Deng [1] and Chaoqun Zeng [2,*]

1 School of Civil Engineering and Transportation, Guangzhou University, Guangzhou 510006, China; zhumiaochang@gzhu.edu.cn (M.Z.); zyw09@e.gzhu.edu.cn (Y.Z.); 2112216336@e.gzhu.edu.cn (H.C.); dengjun@gzhu.edu.cn (J.D.)

2 School of Automotive and Transportation Engineering, Shenzhen Polytechnic University, Shenzhen 518055, China

* Correspondence: zengchaoqun@szpu.edu.cn

Abstract

This study systematically investigates the tensile property degradation of Carbon Fabric-Reinforced Cementitious Matrix (CFRCM) composites under anodic polarization, explicitly comparing the effects of three standard-required electrolyte environments (NACE/ISO). CFRCM specimens were polarized for 20 days at current densities of 200 and 400 mA/m^2 in NaCl, NaOH, and simulated concrete pore solutions. Results reveal that anodic polarization significantly reduces peak tensile strength and post-cracking stiffness, with degradation severity dependent on the electrolyte type (NaCl > NaOH > Pore Solution). Crucially, comparative analysis demonstrates that the strength degradation of carbon fiber bundles embedded in the mortar matrix is more pronounced than that of bare bundles. This work provides essential durability data for CFRCM composites for integrated ICCP-Structural Strengthening systems.

Keywords: CFRCM composites; anodic polarization; tensile properties; tensile stress versus strain curves; tensile strength

1. Introduction

Steel corrosion is recognized as a critical issue threatening the durability and safety of reinforced concrete structures. Corrosion not only leads to reduction in the effective cross-sectional area of the steel reinforcement but, more critically, the volumetric expansion of its products (which can reach 2–6 times the original volume) can induce cracking and spalling of the concrete cover, severely compromising structural capacity [1]. Although traditional strengthening techniques, such as bonding Fiber Reinforced Polymer (FRP), can improve short-term load-bearing capacity [2–5], they are unable to halt the ongoing corrosion process of steel. For preventing corrosion, Impressed Current Cathodic Protection (ICCP) is widely employed as a reliable long-term protection technology [6–13]. It operates by impressing an external current to maintain the steel reinforcement in a stable cathodic state, effectively suppressing its electrochemical corrosion.

To simultaneously achieve long-term anti-corrosion and structural strengthening, an innovative concept integrating these two functions has been proposed, namely ICCP-Structural Strengthening (ICCP-SS) [14–16]. The key advantage of this integrated technique lies in its ability to provide immediate enhancement of load-bearing capacity while also delivering continuous cathodic protection, which is expected to significantly extend the service life of structures and reduce life-cycle costs. The realization of this technique relies on the development of dual-functional materials which serve as structural strengthening elements and impressed current anodes. Carbon Fiber Reinforced Polymer (CFRP) was utilized as a dual-functional material in the early research [17–22]. However, its polymeric resin matrix is susceptible to anodic polarization-induced degradation, and its inherent low electrical conductivity hinders uniform current distribution, limiting the reliability of long-term application.

To address the issues associated with the polymer matrix, Carbon Fabric Reinforced Cementitious Matrix (CFRCM) composites (also termed textile reinforced concrete/mortar), which utilize cementitious materials to replace the resin, have emerged. The cementitious matrix offers excellent compatibility with the concrete substrate and superior performance to FRP under thermal exposure, though the mechanical behavior of FRCM can be compromised under elevated temperatures [23–26].

Extensive research has been conducted to investigate the mechanical properties of CFRCM composites at the material level and their strengthening effectiveness at the member level [27–34]. Materially, CFRCM exhibits typical tensile strain-hardening behavior and multiple cracking under tensile load, and its mechanical constitutive relationship has been well characterized. At the structural member level, studies have confirmed that CFRCM can be effectively used to strengthen beams [28,33,35–37] and columns [38–42].

Nevertheless, when CFRCM composite serves as an anode in an ICCP system subjected to long-term anodic polarization, its durability concerns become prominent. The anodic polarization process may induce electrochemical oxidation of the carbon fibers themselves [43–48], acidification of the cementitious matrix [49–52], and damage at the fiber-matrix interface [43,53,54], leading to the degradation of the mechanical properties of CFRCM composites. Research indicates that with increasing charge density (the product of current density and time), the tensile peak strength and the modulus representing post-cracking stiffness of CFRCM show significant deterioration [55–58]. Microscopic observations have further revealed signs of degradation on the carbon fiber surface after polarization [43,46–48].

It is noteworthy that for ensuring reliability in engineering applications, relevant technical standards (NACE TM0294-2016 [59] and ISO 19097-1:2018 [60]) explicitly require that impressed current anodes for concrete structures must undergo long-term polarization tests in different electrolyte solutions (such as chloride-containing solutions, alkaline solutions, and simulated concrete pore solutions) to assess their durability. However, there is still a lack of systematic research investigating the evolution in the mechanical properties of CFRCM composites after being subjected to anodic polarization in the aforementioned environments.

This study aims to reveal the tensile mechanical properties of CFRCM after anodic polarization in three typical environments: NaCl, NaOH, and simulated pore solution. Through tensile tests on polarized specimens, the failure modes, crack propagation, load versus deformation curves, and stress–strain constitutive relationships were analyzed. Mechanical parameters sensitive to anodic polarization were extracted. Furthermore, by comparing the strength degradation of CFRCM with that of bare carbon fiber bundles, the exacerbating effect of non-uniform degradation caused by the mortar matrix on the fiber degradation process was elucidated. This work provides a critical basis for the durability design and assessment of CFRCM composites in integrated ICCP-SS systems.

2. Materials and Methods

This investigation was designed to reveal the tensile mechanical properties of CFRCM composites after being subjected to anodic polarization in different electrolyte solutions, with a focus on analyzing the failure mode, crack characteristics, peak strength, and

https://doi.org/10.3390/jcs10010016

stress–strain constitutive relationship during tension. The main work included specimen preparation, anodic polarization testing, and tensile testing.

2.1. Raw Materials and Specimen Preparation

CFRCM composites were composed of carbon fiber fabric and mortar matrix. To exclude the influence of the fiber bundle nodes in a bidirectional fabric, the composite used in this study contained only a single longitudinal carbon fiber bundle, and its geometrical configuration followed that documented in the reference [57]. The specimens had a cross-sectional area of 10 mm × 10 mm and a length of 250 mm, with the carbon fiber bundle positioned centrally along the specimen axis (as shown in Figure 1). The carbon fiber bundle was sourced from untwisted continuous fiber roving, with a width of 3 mm. Each bundle contained 12,000 filaments, with the filament diameter being 7 μm. The elastic modulus and tensile strength of the carbon fiber bundle were 230 GPa and 2300 MPa, respectively.

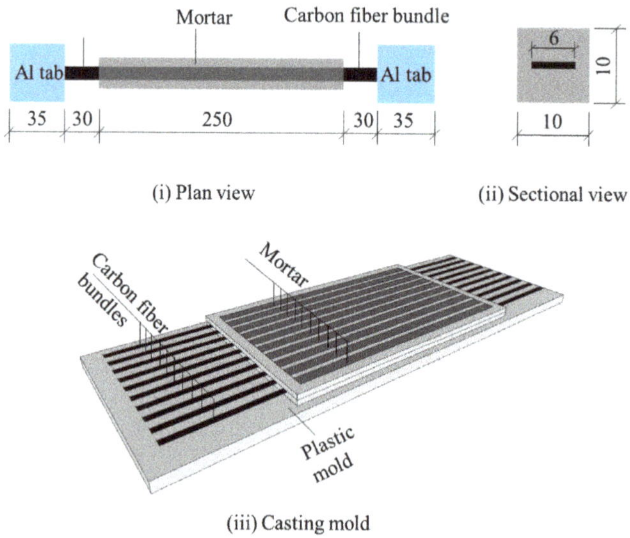

Figure 1. Geometrical details and casting mold for CFRCM composites (unit: mm).

The mortar matrix was prepared using PO 42.5 ordinary Portland cement with a water-to-cement ratio of 0.35. Fine sand with a particle size range of 0–0.18 mm was used as fine aggregate, accounting for 50% of the cement mass. To improve workability, a polycarboxylate superplasticizer, amounting to 0.055% of the cement mass, was added during mixing.

The CFRCM composite specimens were cast using customized plastic molds. Each mold could produce 10 specimens simultaneously (Figure 1). After casting, the molds were immediately sealed with plastic wrap to prevent rapid moisture evaporation which would affect cement hydration. After 24 h, the plastic wrap was removed, and the entire mold containing the CFRCM specimens was transferred to a standard curing room for curing. Besides serving as a mold, this setup also functioned as the electrolytic cell for the anodic polarization device, used to contain the electrolyte solution.

2.2. Implementation of Anodic Polarization

A simulated ICCP setup was used for the galvanostatic anodic polarization tests on CFRCM composites. This setup included the anode (CFRCM composite), the cathode

https://doi.org/10.3390/jcs10010016

(stainless steel plate), the electrolyte solution, the electrical circuit, and a DC power supply (constant current output), as illustrated in Figure 2. To ensure good electrical connectivity, the exposed carbon fiber ends of the CFRCM composite specimen were reliably connected to copper wires using conductive silver paste before being connected to the positive terminal of the power supply. The cathode (stainless steel plate) was connected to the negative terminal.

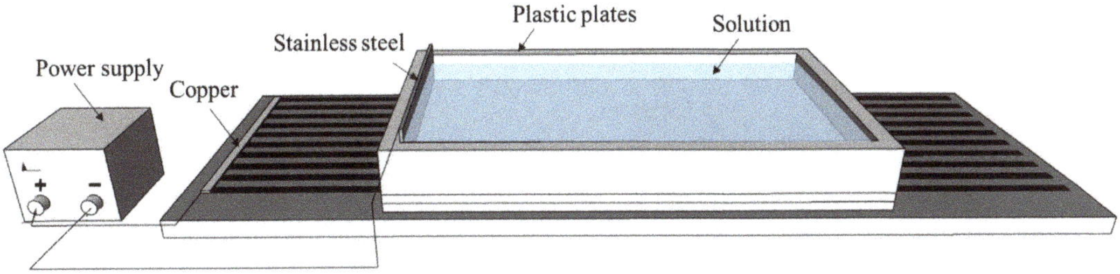

Figure 2. Anodic polarization setup.

Three types of electrolyte solutions were selected based on NACE and ISO standards [59,60]: a 30 g/L NaCl solution, a 40 g/L NaOH solution, and a simulated concrete pore solution (SCP, concentration in weight: 0.20% $Ca(OH)_2$, 3.20% KCl, 1.00% KOH, 2.45% NaOH and 93.15% deionized water), to analyze the effects of different solution environments. Each polarization cell contained ~900 mL of electrolytes which were refreshed twice during polarization. The polarization duration for all specimens was set at 20 days. The anodic current density (relative to the surface area of the carbon fiber bundle) was set at 200 mA/m^2 and 400 mA/m^2. The polarization was conducted indoors with the electrolyte temperature controlled at around 26 °C throughout polarization to prevent significant thermal variations.

The specimens were labeled according to the following convention: Polarization duration (D followed by number of days)-Current density (I followed by current density value)-Solution environment (denoted by letters A, B, C, where A represents 30 g/L NaCl solution, B represents 40 g/L NaOH solution, and C represents the simulated pore solution). For instance, D20-I200-A denotes a specimen polarized in 30 g/L NaCl solution at a current density of 200 mA/m^2 for 20 days. The non-polarized control group specimens were labeled as Ref.

2.3. Tensile Tests

After the polarization period, the CFRCM specimens were removed from the molds and allowed to dry under ambient laboratory conditions. Both ends of the CFRCM composite tensile specimens were reinforced with aluminum alloy plates to prevent slippage of the carbon fiber bundle from the grips during testing. Gripping CF bundles in tension enabled CFRCM specimens to fail by bundle fracture, allowing for comparative assessment of mechanical properties with bare bundles. Note that this gripping method differs from standard characterization tests (e.g., ACI 549.4R [61] or RILEM TC 232-TDT [62]) where the load is transferred through the matrix. While standard methods are preferable for determining design properties, gripping the fiber bundles directly was necessary in this study to ensure fiber rupture, thereby allowing for a direct quantification of the degradation of carbon fiber bundles within the matrix.

The tensile tests were conducted on a universal testing machine with a 10 kN load cell, as shown in Figure 3. The tests were displacement-controlled, with a loading rate of 2 mm/min. The load applied to the specimen was measured by the machine's built-in load

https://doi.org/10.3390/jcs10010016

cell. After excluding machine compliance following toe compensation (ASTM D638 [63]), the tensile deformation was calculated by deducting the elastic deformation of the exposed fiber bundle from the total machine crosshead displacement. Both load and deformation data were synchronously recorded by a data acquisition system.

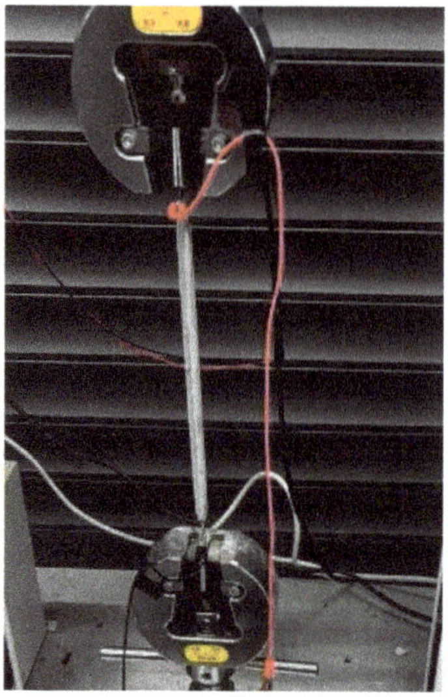

Figure 3. Tensile tests for CFRCM composite specimens.

3. Results

3.1. Failure Modes

After tensile testing, the CFRCM specimens primarily exhibited two failure modes, as shown in Figure 4. The control specimens (Ref) showed multiple cracking in the mortar matrix during tension. At the peak load, failure occurred due to the rupture of the exposed carbon fiber bundle outside the mortar matrix, as shown in Figure 4a. After anodic polarization, the specimens still developed multiple cracks in the mortar matrix during tension, but the average crack spacing was noticeably increased. At the peak load, the location of the carbon fiber bundle rupture shifted to within a mortar crack (Figure 4b). This indicates that the anodic polarization process may have caused degradation at the carbon fiber-mortar interface or within the carbon fibers themselves, thereby changing the failure location of the specimens.

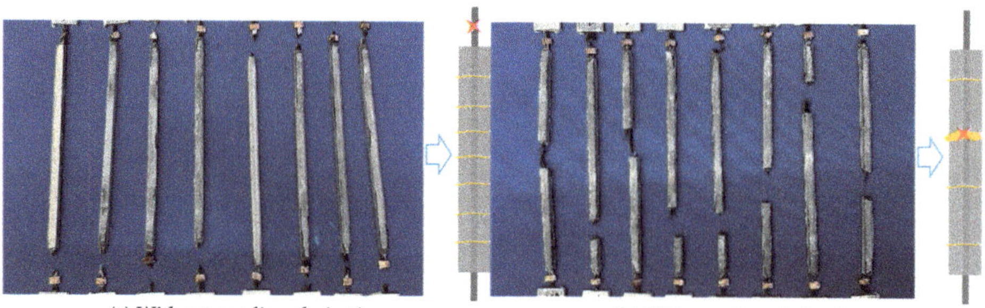

(**a**) Without anodic polarization (**b**) With anodic polarization

Figure 4. Failure patterns for CFRCM composites without (**a**) and with (**b**) anodic polarization.

https://doi.org/10.3390/jcs10010016

3.2. Load–Displacement Curves

The typical load–displacement curves of the CFRCM composite specimens are depicted in Figure 5. All specimens exhibited characteristic strain-hardening behavior. Their mechanical response can be divided into three stages: the initial linear-elastic uncracked stage, the multiple cracking stage, and the crack-widening stage. In the initial stage, the load and deformation showed a linear relationship, as the stress was primarily carried by the mortar matrix. Once the load reached the tensile strength of the mortar, matrix cracking initiated, leading to a reduction in the slope of the curve accompanied by fluctuations, marking the beginning of the multiple cracking stage where the number of cracks increased and the crack spacing decreased. This was followed by the crack-widening stage, where the curve ascended approximately linearly, but with a slope lower than the initial elastic stage, until the rupture of the carbon fiber bundle resulted in final failure. Anodic polarization significantly affected the mechanical behavior of the specimens in the crack-widening stage, manifesting as a decrease in the slope of the curve in this stage and a reduction in the peak load. This phenomenon can be attributed to the degradation of the carbon fibers themselves and the fiber-matrix interface induced by anodic polarization [43,46,48].

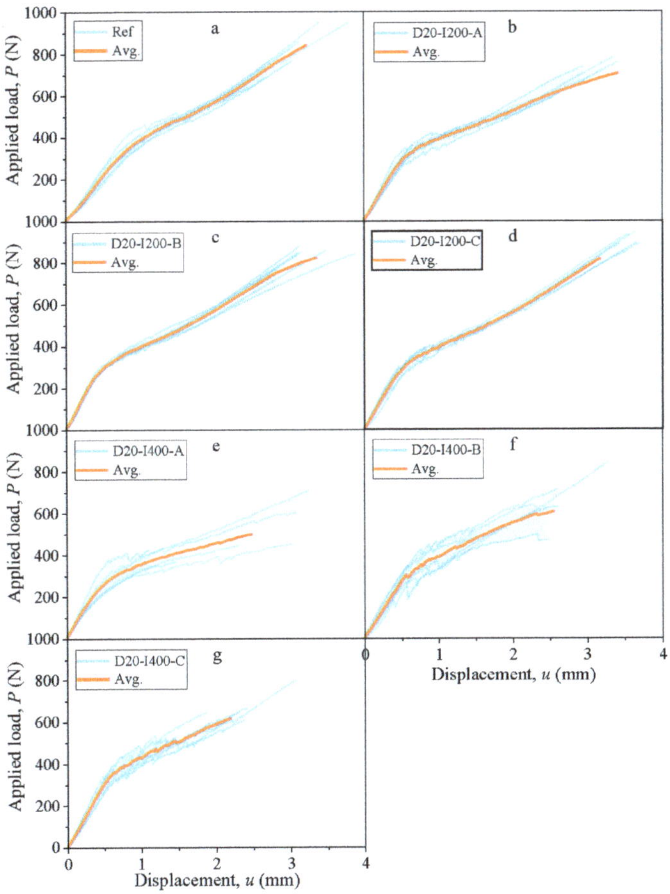

Figure 5. Applied load versus displacement curves.

https://doi.org/10.3390/jcs10010016

3.3. The Number of Cracks Within Mortar

The number of cracks on the surface of the CFRCM specimens after tensile testing was counted. The average values and standard deviations for each polarization condition were calculated. The average crack numbers for the CFRCM specimens are plotted in Figure 6. Compared to the control group (average number of 16), the number of cracks in the anodically polarized specimens decreased with increasing current density and was influenced by the type of electrolyte solution. After 20 days of polarization at 400 mA/m^2, the average number of cracks decreased to 9.75, 10.2, and 8 for specimens polarized in solutions A, B, and C, respectively.

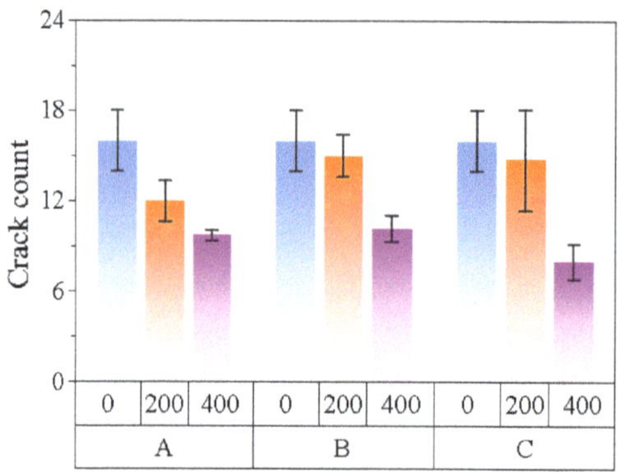

Figure 6. The number of cracks in the CFRCM composite matrix.

The reduction in the number of cracks confirms the degradation effect induced by anodic polarization. The mechanism of crack spacing formation is related to stress transfer: a crack forms when the stress in the mortar reaches its tensile strength, transferring the load at the crack to the carbon fiber bundle. Through the interfacial bond stress between the fiber bundle and the matrix, the load is gradually transferred back to the mortar matrix until the stress in the mortar reaches its tensile strength again at a sufficient distance, forming a new crack. Anodic polarization weakens the interfacial bond strength, leading to an increase in the required bond transfer length (half the crack spacing) to complete the stress transfer, consequently resulting in a reduced number of cracks in the specimen.

3.4. Tensile Stress–Strain Constitutive Relationship

The tensile stress in the CFRCM composite was calculated by dividing the applied load by the cross-sectional area of the carbon fiber bundle. The average strain was determined by dividing the tensile deformation of the specimen by its gauge length. The established tensile stress–strain constitutive relationship for CFRCM is shown in Figure 7, sharing a similar shape with the load–displacement curves. Given the relatively brief duration of the multiple cracking stage, the constitutive relationship can be simplified into a bilinear model, comprising the elastic uncracked stage and the post-cracking stage. The effect of anodic polarization was primarily reflected in the changes in the modulus of the post-cracking stage and the peak strength, as illustrated in Figure 8.

https://doi.org/10.3390/jcs10010016

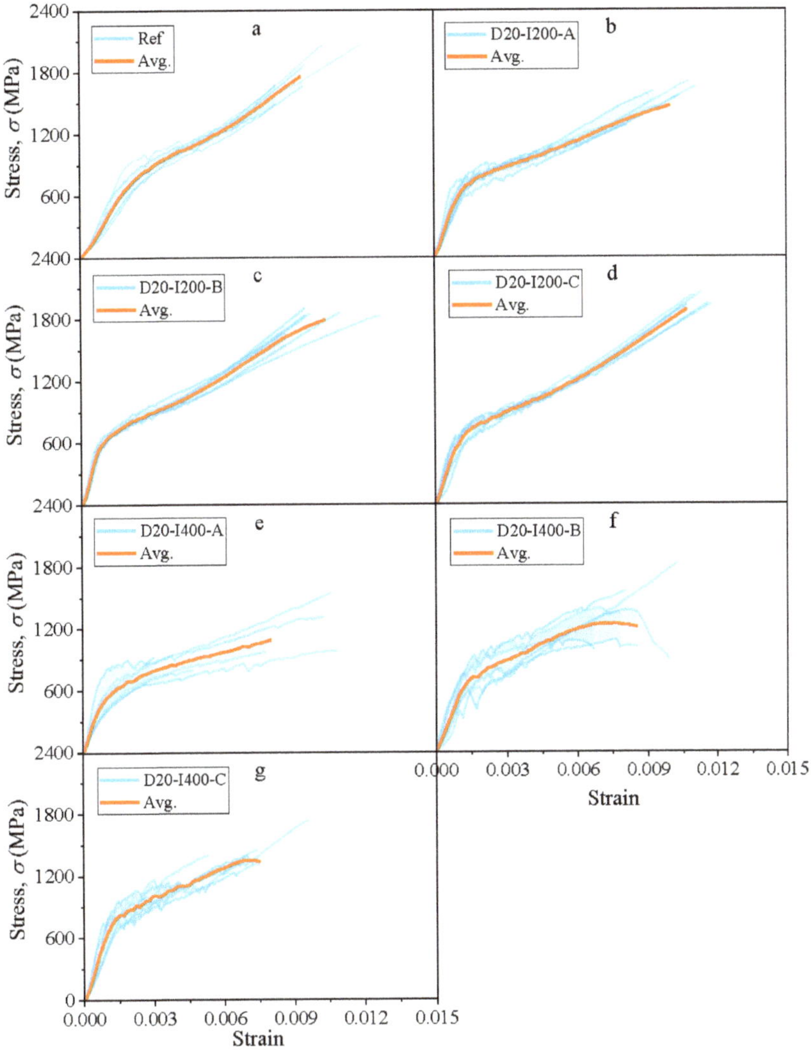

Figure 7. Tensile stress versus strain curves.

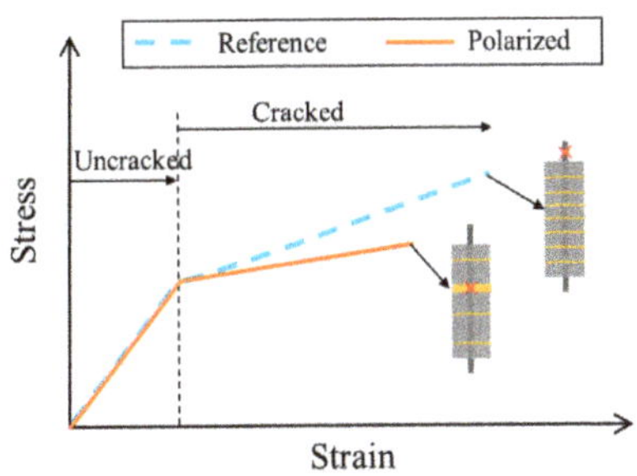

Figure 8. Tensile stress versus strain curves for CFRCM composites with and without anodic polarization.

https://doi.org/10.3390/jcs10010016

To quantify the effect of anodic polarization, the elastic modulus (slope of the curve) in the post-cracking stage and the peak strength of the polarized specimens were normalized against the corresponding values of the control specimens. Charge density (the product of current density and time) was introduced as a comprehensive indicator of the polarization level. The relationship between the normalized mechanical properties and the charge density is shown in Figure 9. At a charge density of 691,200 C/m^2, the modulus in the post-cracking stage decreased by 54.5%, 26.8%, and 15.1% for specimens polarized in solution A (NaCl), B (NaOH), and C (SCP), respectively. Correspondingly, the peak tensile strength was reduced by 38.7%, 22.0%, and 19.2%. This indicates that the NaCl solution environment had the most significant degrading effect on the mechanical properties of CFRCM, while the effects in the NaOH and simulated pore solutions were relatively weaker.

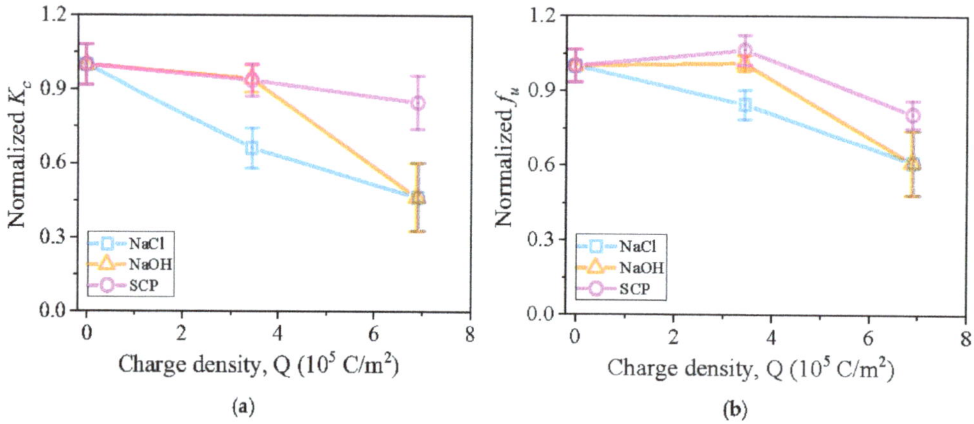

Figure 9. Effects of anodic polarization on mechanical properties of CFRCM composites: (**a**) Elastic modulus at the cracked stage; (**b**) Tensile strength.

3.5. Comparison of Tensile Strength: Bare Carbon Fiber Bundles vs. CFRCM-Embedded Bundles

Since all CFRCM specimens ultimately failed due to the rupture of the carbon fiber bundle, their tensile strength directly reflects the residual strength of the fiber bundle. Figure 10 compares the normalized tensile strength of bare carbon fiber bundles and the bundles embedded within CFRCM composites after polarization under various conditions. The results demonstrate that for any given electrolyte solution, the degree of strength degradation of the carbon fiber bundles within the CFRCM composite specimens was more severe than that of the bare carbon fiber bundles. A plausible explanation is that within the CFRCM, the distribution of current within the carbon fiber bundle, composed of tens of thousands of filaments, may become non-uniform due to the presence of the mortar matrix. This non-uniformity could lead to localized areas of high current density, thereby exacerbating the non-uniform degradation of the fiber bundle. Consequently, the non-uniform polarization effect induced by the mortar matrix must be considered when evaluating the impact of anodic polarization on the overall performance of CFRCM composites.

https://doi.org/10.3390/jcs10010016

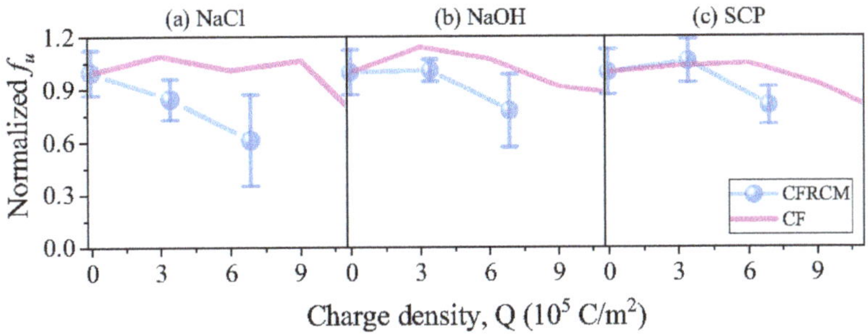

Figure 10. Comparisons in normalized tensile strength between bare carbon fiber bundles and CFRCM composites subjected to anodic polarization.

4. Conclusions

Based on the tensile tests conducted on carbon fabric-reinforced cementitious matrix composites subjected to anodic polarization in different electrolyte solutions, the following main conclusions were drawn:

(1) The anodic polarization treatment altered the tensile failure mode of the CFRCM specimens. In the non-polarized control specimens, the rupture of the exposed carbon fiber bundle occurred outside the mortar matrix. In contrast, after anodic polarization, the rupture of the carbon fiber bundle consistently occurred within a crack inside the mortar matrix. This indicates that the anodic polarization process caused degradation at the carbon fiber-mortar interface and within the carbon fibers themselves, prompting a shift in the location of fiber rupture.

(2) Compared to the control group, the polarized specimens exhibited a reduced number of cracks and an increased average crack spacing. This confirms that anodic polarization weakened the interfacial bond strength between the fibers and the mortar matrix.

(3) All specimens exhibited typical strain-hardening behavior. However, the slope of the load-deformation curve in the crack-widening stage was noticeably reduced after anodic polarization. This indicates that anodic polarization not only reduced the peak strength of the material but also impaired its stiffness during the primary service stage (post-cracking stage).

(4) The degrading effect of anodic polarization on the post-cracking modulus and peak strength of CFRCM composites was positively correlated with the charge density (the product of current density and polarization time) and was significantly modulated by the type of electrolyte solution. At a charge density of 691,200 C/m^2, the property degradation was most pronounced in the NaCl solution (A: 54.5% reduction in modulus, 38.7% reduction in strength), while it was least significant in the simulated pore solution (C: 15.1% reduction in modulus, 19.2% reduction in strength).

(5) Comparison with bare carbon fiber bundles directly exposed to the solutions revealed that the strength degradation of the bundles embedded within CFRCM composites was more severe. This highlights a critical issue caused by the mortar matrix: non-uniform current distribution. This non-uniformity induces more significant localized damage, thereby exacerbating the overall degradation of mechanical performance. Therefore, the non-uniform polarization effect resulting from the matrix must be considered in the durability assessment of CFRCM composite for long-term applications.

https://doi.org/10.3390/jcs10010016

Author Contributions: Conceptualization, M.Z. and C.Z.; methodology, Y.Z.; software, J.D.; validation, M.Z., J.D. and C.Z.; formal analysis, H.C.; investigation, Y.Z. and H.C.; resources, J.D.; data curation, M.Z.; writing—original draft preparation, M.Z.; writing—review and editing, C.Z.; visualization, M.Z.; supervision, J.D. and C.Z.; project administration, M.Z.; funding acquisition, M.Z. and C.Z. All authors have read and agreed to the published version of the manuscript.

Funding: This research was funded by the National Natural Science Foundation of China (52208277 and 52208308).

Institutional Review Board Statement: Not applicable. This study did not involve humans or animals.

Informed Consent Statement: Not applicable. This study did not involve humans or animals.

Data Availability Statement: The raw data supporting the conclusions of this article will be made available by the authors on request.

Conflicts of Interest: The authors declare no conflicts of interest.

References

1. Broomfield, J.P. *Corrosion of Steel in Concrete*; CRC Press: London, UK, 2022; ISBN 9781003223016.
2. Duan, L.; Wang, C.; Brühwiler, E.; Wang, Q. Experiments on the Flexural Behavior of Full-Scale PC Box Girders with Service Damage Strengthened by Prestressed CFRP Plates. *Compos. Struct.* **2023**, *312*, 116864. [CrossRef]
3. Kotynia, R.; Oller, E.; Marí, A.; Kaszubska, M. Efficiency of Shear Strengthening of RC Beams with Externally Bonded FRP Materials—State-of-the-Art in the Experimental Tests. *Compos. Struct.* **2021**, *267*, 113891. [CrossRef]
4. Hu, W.; Li, Y.; Yuan, H. Review of Experimental Studies on Application of FRP for Strengthening of Bridge Structures. *Adv. Mater. Sci. Eng.* **2020**, *2020*, 1–21. [CrossRef]
5. Naser, M.Z.; Hawileh, R.A.; Abdalla, J.A. Fiber-Reinforced Polymer Composites in Strengthening Reinforced Concrete Structures: A Critical Review. *Eng. Struct.* **2019**, *198*, 109542. [CrossRef]
6. Byrne, A.; Norton, B.; Holmes, N. State-of-the-Art Review of Cathodic Protection for Reinforced Concrete Structures. *Mag. Concr. Res.* **2016**, *68*, 664–677. [CrossRef]
7. Bertolini, L.; Pedeferri, P.; Redaelli, E.; Pastore, T. Repassivation of Steel in Carbonated Concrete Induced by Cathodic Protection. *Mater. Corros.* **2003**, *54*, 163–175. [CrossRef]
8. Bhuiyan, S. Effectiveness of Impressed Current Cathodic Protection in Concrete Following Current Interruption. Master's Thesis, University of Sheffield, Sheffield, UK, 2015.
9. Polder, R.; Peelen, W. Cathodic Protection of Steel in Concrete—Experience and Overview of 30 Years Application. *MATEC Web Conf.* **2018**, *199*, 01002. [CrossRef]
10. Cheaitani, A. Review of Cathodic Protection Systems for Concrete Structures in Australia. In Proceedings of the CORROSION 2017, New Orleans, LA, USA, 26–30 March 2017; pp. 3–4.
11. Pedeferri, P. Cathodic Protection and Cathodic Prevention. *Constr. Build. Mater.* **1996**, *10*, 391–402. [CrossRef]
12. Mietz, J.; Fischer, J.; Isecke, B. Cathodic Protection of Steel-Reinforced Concrete Structures—Results from 15 Years' Experience. *Mater. Perform.* **2001**, *40*, 22–26.
13. Polder, R.B.; Peelen, W.H.A. *Service Life Aspects of Cathodic Protection of Concrete Structures*; Taylor & Francis Group: Abingdon, UK, 2008.
14. Zhu, J.; Su, M.-N.; Huang, J.; Ueda, T.; Xing, F. The ICCP-SS Technique for Retrofitting Reinforced Concrete Compressive Members Subjected to Corrosion. *Constr. Build. Mater.* **2018**, *167*, 669–679. [CrossRef]
15. Su, M.; Liang, H.; Wei, L.; Ueda, T.; Xing, F.; Zhu, J.; Zeng, Z.W. Experimental Investigation on the ICCP-SS Technique for Sea-Sand RC Beams. In Proceedings of the Sixth International Conference on Durability of Concrete Structures, Leeds, UK, 18–20 July 2018.
16. Su, M.; Wei, L.; Zeng, Z.; Ueda, T.; Xing, F.; Zhu, J.-H. A Solution for Sea-Sand Reinforced Concrete Beams. *Constr. Build. Mater.* **2019**, *204*, 586–596. [CrossRef]
17. Sun, H.; Wei, L.; Zhu, M.; Han, N.; Zhu, J.; Xing, F. Corrosion Behavior of Carbon Fiber Reinforced Polymer Anode in Simulated Impressed Current Cathodic Protection System with 3% NaCl Solution. *Constr. Build. Mater.* **2016**, *112*, 538–546. [CrossRef]
18. Zhu, J.; Guo, G.; Wei, L.; Zhu, M.; Chen, X. Dual Function Behavior of Carbon Fiber-Reinforced Polymer in Simulated Pore Solution. *Materials* **2016**, *9*, 103. [CrossRef] [PubMed]
19. Zhu, J.-H.; Wei, L.; Moahmoud, H.; Redaelli, E.; Xing, F.; Bertolini, L. Investigation on CFRP as Dual-Functional Material in Chloride-Contaminated Solutions. *Constr. Build. Mater.* **2017**, *151*, 127–137. [CrossRef]

https://doi.org/10.3390/jcs10010016

20. Zhu, J.; Wei, L.; Zhu, M.; Sun, H.; Tang, L.; Xing, F. Polarization Induced Deterioration of Reinforced Concrete with CFRP Anode. *Materials* **2015**, *8*, 4316–4331. [CrossRef]
21. Wei, L.; Zhu, J.-H.; Dong, Z.; Liu, J.; Liu, W.; Su, M.; Xing, F. Anodic and Mechanical Behavior of Carbon Fiber Reinforced Polymer as a Dual-Functional Material in Chloride-Contaminated Concrete. *Materials* **2020**, *13*, 222. [CrossRef]
22. Sun, H.; Memon, S.A.; Gu, Y.; Zhu, M.; Zhu, J.; Xing, F. Degradation of Carbon Fiber Reinforced Polymer from Cathodic Protection Process on Exposure to NaOH and Simulated Pore Water Solutions. *Mater. Struct.* **2016**, *49*, 5273–5283. [CrossRef]
23. Mazzuca, P.; Ombres, L.; Guglielmi, M.; Verre, S. Residual Mechanical Properties of PBO FRCM Composites after Elevated Temperature Exposure: Experimental and Comparative Analysis. *J. Mater. Civ. Eng.* **2023**, *35*, 04023383. [CrossRef]
24. Mazzuca, P.; Micieli, A.; Campolongo, F.; Ombres, L. Influence of Thermal Exposure Scenarios on the Residual Mechanical Properties of a Cement-Based Composite System. *Constr. Build. Mater.* **2025**, *466*, 140304. [CrossRef]
25. Ombres, L.; Mazzuca, P.; Micieli, A.; Candamano, S.; Campolongo, F. FRCM–Masonry Joints at High Temperature: Residual Bond Capacity. *J. Mater. Civ. Eng.* **2025**, *37*, 04025012. [CrossRef]
26. Asghari, H.; Omeman, Z.; Noel, M.; Hajiloo, H. Tensile Properties of Carbon Fabric-Reinforced Cementitious Matrix (FRCM) at High Temperatures. *Structures* **2023**, *55*, 85–96. [CrossRef]
27. Raoof, S.M.; Koutas, L.N.; Bournas, D.A. Bond between Textile-Reinforced Mortar (TRM) and Concrete Substrates: Experimental Investigation. *Compos. Part B* **2016**, *98*, 350–361. [CrossRef]
28. Bencardino, F.; Carloni, C.; Condello, A.; Focacci, F.; Napoli, A.; Realfonzo, R. Flexural Behaviour of RC Members Strengthened with FRCM: State-of-the-Art and Predictive Formulas. *Compos. Part B* **2018**, *148*, 132–148. [CrossRef]
29. Younis, A.; Ebead, U. Bond Characteristics of Different FRCM Systems. *Constr. Build. Mater.* **2018**, *175*, 610–620. [CrossRef]
30. D'Antino, T.; Papanicolaou, C. Mechanical Characterization of Textile Reinforced Inorganic-Matrix Composites. *Compos. Part B Eng.* **2017**, *127*, 78–91. [CrossRef]
31. Arboleda, D. Fabric Reinforced Cementitious Matrix (FRCM) Composites for Infrastructure Strengthening and Rehabilitation: Characterization Methods. Ph.D. Thesis, University of Miami, Coral Gables, FL, USA, 2014.
32. Al-Lami, K.; D'Antino, T.; Colombi, P. Durability of Fabric-Reinforced Cementitious Matrix (FRCM) Composites: A Review. *Appl. Sci.* **2020**, *10*, 1714. [CrossRef]
33. Daneshvar, K.; Moradi, M.J.; Roshan, N.; Noel, M.; Hajiloo, H. Enhancing Flexural and Shear Capacities of RC T-Beams with FRCM Incorporating a Full FRCM-Concrete Bond. *Constr. Build. Mater.* **2025**, *471*, 140687. [CrossRef]
34. Irandegani, M.A.; Zhang, D.; Shadabfar, M.; Kontoni, D.P.N.; Iqbal, M. Failure Modes of RC Structural Elements and Masonry Members Retrofitted with Fabric-Reinforced Cementitious Matrix (FRCM) System: A Review. *Buildings* **2022**, *12*, 653. [CrossRef]
35. Liu, X.; Figueredo, G.P.; Gordon, G.S.D.; Thermou, G.E. Data-Driven Shear Strength Prediction of RC Beams Strengthened with FRCM Jackets Using Machine Learning Approach. *Eng. Struct.* **2025**, *325*, 119485. [CrossRef]
36. Sheng, J.; Dou, G.; Bi, X.; Zhu, J.; Yu, Z. Experimental Study on the Flexural Fatigue Performance of TRC-Strengthened Corroded RC Beams. *J. Build. Eng.* **2025**, *110*, 113065. [CrossRef]
37. Bertolli, V.; Sneed, L.H.; Focacci, F.; D'Antino, T. Shear Strengthening of RC Beams with U-Wrapped FRCM Composites: State of the Art and Assessment of Available Analytical Models. *J. Compos. Constr.* **2025**, *29*, 04024091. [CrossRef]
38. Zhu, J.-H.; Wang, Z.; Su, M.; Ueda, T.; Xing, F. C-FRCM Jacket Confinement for RC Columns under Impressed Current Cathodic Protection. *J. Compos. Constr.* **2020**, *24*, 4020001. [CrossRef]
39. Elnassar, Z.; Abed, F.; Refai, A.E.; El-Maaddawy, T. FRCM Confinement of Concrete Columns: A Review of Strength and Ductility Enhancements. *Compos. Struct.* **2025**, *370*, 119389. [CrossRef]
40. Jawdhari, A.; Adheem, A.H.; Kadhim, M.M.A. Parametric 3D Finite Element Analysis of FRCM-Confined RC Columns under Eccentric Loading. *Eng. Struct.* **2020**, *212*, 110504. [CrossRef]
41. Kumar, P.; Arora, H.C.; Chidambaram, R.S.; Kumar, A. Prediction of Confined Compressive Strength of Concrete Column Strengthened with FRCM Composites. *Struct. Concr.* **2025**, *26*, 879–908. [CrossRef]
42. Faleschini, F.; Zanini, M.A.; Hofer, L.; Pellegrino, C. Experimental Behavior of Reinforced Concrete Columns Confined with Carbon-FRCM Composites. *Constr. Build. Mater.* **2020**, *243*, 118296. [CrossRef]
43. Zhu, M.; Zhu, J.; Ueda, T.; Matsumoto, K. Degradation Behavior of Multifunctional Carbon Fabric-Reinforced Cementitious Matrix Composites under Anodic Polarization. *Constr. Build. Mater.* **2022**, *341*, 127751. [CrossRef]
44. Chen, X.; Zhang, C.; Song, G.; Zheng, D.; Guo, Y.; Huang, X. Electrochemical Activity and Damage of Single Carbon Fiber. *Materials* **2021**, *14*, 1758. [CrossRef]
45. Sun, Y.; Lu, Y.; Yang, C. Stripping Mechanism of PAN-Based Carbon Fiber during Anodic Oxidation in NaOH Electrolyte. *Appl. Surf. Sci.* **2019**, *486*, 128–136. [CrossRef]
46. Zhu, J.-H.; Li, Q.; Pei, C.; Yu, H.; Xing, F. Evolution Mechanism of Carbon Fiber Anode Properties for Functionalized Applications: Impressed Current Cathodic Protection and Structural Strengthening. *Engineering* **2025**, *in press*. [CrossRef]
47. Pei, C.; Yu, H.; Zhu, J.; Xing, F. Efficient Multifunctional Modification of Commercial Carbon Fiber Through Tailored Carbon Layer Structure. *Engineering* **2024**, *55*, 191–203. [CrossRef]

https://doi.org/10.3390/jcs10010016

48. Yu, H.; Li, Q.; Zhu, J.-H.; Xing, F. Anodic Degradation Behaviour of Carbon Fibre in CFRP at High-Chloride and -Alkali Condition. *Constr. Build. Mater.* **2024**, *417*, 135241. [CrossRef]
49. Zhu, J.H.; Wu, X.Y.; Mohamed, I.M.A.; Xing, F. Electrochemical and Microstructural Evaluation of Acidification Damage Induced by Impressed Current Cathodic Protection after Incorporating a Hydroxy Activated-Mg/Al-Double Oxide in the External Anode Mortar. *Constr. Build. Mater.* **2021**, *309*, 125116. [CrossRef]
50. Koster, T.; Peelen, W.; Larbi, J.; de Rooij, M.; Polder, R. Numerical Model of $Ca(OH)_2$ Transport in Concrete Due to Electrical Currents. *Mater. Corros.* **2010**, *61*, 518–523. [CrossRef]
51. Guo, W.; Hu, J.; Ma, Y.; Huang, H.; Yin, S.; Wei, J.; Yu, Q. The Application of Novel Lightweight Functional Aggregates on the Mitigation of Acidification Damage in the External Anode Mortar during Cathodic Protection for Reinforced Concrete. *Corros. Sci.* **2020**, *165*, 108366. [CrossRef]
52. Zeng, C.; Qin, S.; Deng, Z.; Zhu, M. In Situ Monitoring of Anodic Acidification Process Using 3D μ-XCT Method. *Materials* **2024**, *17*, 5662. [CrossRef] [PubMed]
53. Tao, Y.; Liu, J.; Zhang, D.; Xie, Z.; Hong, J. Effect of ICCP on Bond Performance and Piezoresistive Effect of Carbon Fiber Bundles in Cementitious Matrix. *Cem. Concr. Compos.* **2024**, *152*, 105645. [CrossRef]
54. Feng, R.; Zhang, J.; Zhu, J.-H.; Xing, F. Experimental Study on Interface Bonding Fatigue Behavior of C-FRCM Composites in ICCP. *Constr. Build. Mater.* **2020**, *259*, 120655. [CrossRef]
55. Feng, R.; Zhang, J.; Zhu, J.-H.; Xing, F. Experimental Study on the Behavior of Carbon-Fabric Reinforced Cementitious Matrix Composites in Impressed Current Cathodic Protection. *Constr. Build. Mater.* **2020**, *264*, 120655. [CrossRef]
56. Wei, L.L.; Zhu, J.-H.; Ueda, T.; Su, M.; Liu, J.; Liu, W.; Tang, L.-P.; Xing, F. Tensile Behaviour of Carbon Fabric Reinforced Cementitious Matrix Composites as Both Strengthening and Anode Materials. *Compos. Struct.* **2020**, *234*, 111675. [CrossRef]
57. Liu, J.; Zhang, D.; Tao, Y.; Xie, Z.; Yi, B. Effect of ICCP on Tensile Performance and Piezoresistive Effect of CFRCM Plates. *Eng. Struct.* **2024**, *313*, 118317. [CrossRef]
58. Wei, L.; Zheng, Y.; Tian, J.; Shen, H.; Zhu, J. Effects of Hybrid Fibers and Anodic Polarization on Mechanical Performance of Carbon Fabric Reinforced Cementitious Matrix (C-FRCM). *Eng. Struct.* **2025**, *336*, 120302. [CrossRef]
59. *NACE TM0294-2016*; Testing of Embeddable Impressed Current Anodes for Use in Cathodic Protection of Atmospherically Exposed Steel-Reinforced Concrete. AMPP: Houston, TX, USA, 2016.
60. *ISO 19097-1:2018*; Accelerated Life Test Method of Mixed Metal Oxide Anodes for Cathodic Protection—Part 1: Application in Concrete. ISO: Geneva, Switzerland, 2018.
61. *ACI 549.4R-13*; Guide to Design and Construction of Externally Bonded Fabric-Reinforced Cementitious Matrix (FRCM) Systems for Repair and Strengthening Concrete and Masonry Structures. American Concrete Institute: Farmington Hills, MI, USA, 2013; p. 69.
62. Brameshuber, W. Recommendation of RILEM TC 232-TDT: Test Methods and Design of Textile Reinforced Concrete. *Mater. Struct.* **2016**, *49*, 4923–4927. [CrossRef]
63. *ASTM D638-10*; Test Method for Tensile Properties of Plastics. ASTM Int.: West Conshohocken, PA, USA, 2010; pp. 1–16.

Disclaimer/Publisher's Note: The statements, opinions and data contained in all publications are solely those of the individual author(s) and contributor(s) and not of MDPI and/or the editor(s). MDPI and/or the editor(s) disclaim responsibility for any injury to people or property resulting from any ideas, methods, instructions or products referred to in the content.

https://doi.org/10.3390/jcs10010016

Journal of
Composites Science

MDPI

Article

Mechanical and Microstructural Performance of Fly Ash-Based Geopolymer Mortar Activated by Silica Fume-Derived Sodium Silicate

Navid Pourdolat, Prakriti Raizada and Rishi Gupta *

Department of Civil Engineering, University of Victoria, Victoria, BC V8P 5C2, Canada; navidpt@outlook.com (N.P.); praizada@uvic.ca (P.R.)
* Correspondence: guptar@uvic.ca

Abstract

The construction industry faces growing pressure to adopt sustainable materials due to the high CO_2 emissions associated with ordinary Portland cement (OPC) production. Geopolymers synthesized from industrial by-products such as fly ash offer a promising low-carbon alternative. However, the extensive use of commercial sodium silicate (SSC) as an activator remains constrained by its high cost and energy-intensive manufacturing. This study investigates a silica fume-derived sodium silicate alternative (SSA) combined with NaOH as a more sustainable activator for fly ash-based geopolymer mortar. Mortars were prepared with alkali activator-to-precursor (AA/P) ratios of 0.7 and 0.5 and cured at 65 °C and 80 °C. SSA-based mixes exhibited comparable flowability to SSC-based mortars, with slightly longer setting times making them favorable for placement. Mechanical tests showed the superior performance of SSA systems, with AS0.7-65 achieving the highest compressive strength and AS0.7-80 demonstrating greater flexural and tensile strength. Microstructural analyses (SEM, EDX, ATR-FTIR) revealed denser matrices and enhanced sodium aluminosilicate hydrate (N-A-S-H) and calcium-rich N(C)-A-S-H gel formation. Economic assessment indicated approximately 30% cost reduction and a modest (~2%) decrease in CO_2 emissions. These findings highlight SSA as a technically viable and sustainable activator for next-generation geopolymer construction.

Keywords: geopolymer; fly ash; silica fume–derived activator; sustainability; microstructure; mechanical properties

1. Introduction

Traditional construction materials, particularly ordinary Portland cement (OPC), are associated with high CO_2 emissions that raise significant environmental concerns. Concrete is the second most utilized substance globally after water, with annual consumption exceeding 30 billion tons. The production of OPC emits approximately 630–800 kg of CO_2 per ton, depending on the fuel source and the decarbonation of limestone required to form clinker phases such as C_3S and C_2S [1]. Consequently, the search for alternative binders that avoid the energy-intensive clinkerization process has become a major research focus [2].

Fly ash-based geopolymers have emerged as one of the most promising alternatives, using industrial by-products such as fly ash to reduce energy demand and minimize carbon footprints. These aluminosilicate binders are formed by alkaline activation, producing three-dimensional inorganic polymer networks [3]. Geopolymers are valued for their high mechanical strength, low shrinkage, thermal stability, and fire resistance [4]. More recently, they have also gained attention as specialized binders for 3D printing, due to their rapid setting and early strength development [5]. However, widespread implementation has been hindered by the reliance on commercial alkaline activators, most notably sodium silicate (SSC). These activators are not only costly but also associated with additional CO_2 emissions that offset some of the sustainability benefits of geopolymers.

Rapid geopolymerization reactions take place when natural or artificial aluminosilicate precursors such as Fly ash, Ground Granulated blast Furnace Slag (GGBS), red mud, volcanic ash, calcined clays, etc., react with alkaline solutions such as NaOH (sodium hydroxide), KOH (potassium hydroxide), or a mixture of these hydroxide solutions and their corresponding silicate solutions [6]. Hydroxide activators promote crystallinity in geopolymers by dissolving aluminosilicate precursors, thereby increasing the availability of aluminum and silicon ions necessary for the formation of crystalline structures [7,8]. Geopolymers activated with NaOH demonstrate enhanced resistance to acids and sulfates and improved thermal stability and mechanical strength [9]. Silicate-based activators, such as sodium silicate (Na_2SiO_3) and potassium silicate (K_2SiO_3), introduce additional silicate species that create a more cross-linked amorphous network, leading to higher ductility and improved chemical resistance [10]. Binary combinations of hydroxides and silicates balance these effects, yielding denser and more homogeneous microstructures [11]. Sodium-based activators are generally more effective than potassium-based ones, producing higher compressive strength and a refined pore structure [12]. However, sodium silicate—the most widely used activator—has critical drawbacks.

Alkaline activators for geopolymers are commonly classified as hydroxide- or silicate-based systems using sodium or potassium cations. Hydroxide activators promote rapid precursor dissolution but often lead to lower polymerization and strength, whereas silicate-based activators provide additional reactive silica, resulting in more cross-linked gel networks and improved mechanical performance. Sodium-based activators generally outperform potassium-based systems due to their smaller ionic radius and more efficient gel formation. However, commercial sodium silicate is energy-intensive and costly to produce, motivating the development of waste-derived alternatives such as silica fume-derived sodium silicate. In the conventional fusion process, sodium carbonate (Na_2CO_3) and quartz (SiO_2) are calcined at 1400–1500 °C [13,14], then dissolved in water and filtered to form the product [15]. Hydrothermal synthesis requires lower temperatures (225–245 °C) but involves high-pressure autoclaves (27–32 bar), reducing overall energy efficiency [16]. Both methods result in significant embodied energy and CO_2 emissions, undermining the sustainability of sodium silicate-activated systems. Recent cost assessments of alkali-activated concretes (AACs) suggest large variability, with costs ranging from 70.67 to 114.78 CAD/m^3, in some cases even higher than OPC-based concrete [17]. Thus, replacing or reducing commercial sodium silicate in geopolymers is essential for advancing economic and environmental viability.

In response, researchers have explored alternative sources of reactive silica to synthesize sodium silicate or sodium silicate alternatives (SSA). Rice husk ash (RHA) and silica fume (SF), both abundant industrial by-products, have shown promise. Tong et al. [14] used RHA to synthesize sodium silicate for alkali-activated binders. Sun et al. [18] evaluated slag activated with SF-derived sodium silicate, reporting improved rheology due to the "ball-bearing" effect of undissolved particles. Cheng et al. [19] demonstrated that silica fume–derived activators reduced production costs and CO_2 emissions when used in fly ash binders. Similarly, Billong et al. [6,20], Oti et al. [21] and Adeleke et al. [22] confirmed that RHA- and SF-derived sodium silicate can achieve comparable or superior strength and durability relative to commercial sodium silicate. These findings suggest that SSA can

https://doi.org/10.3390/jcs10010022

reduce reliance on conventional SSC, but systematic evaluations integrating mechanical performance, microstructural analysis, and sustainability metrics remain limited.

Microstructural characterization is critical for understanding geopolymer performance and durability. Techniques such as scanning electron microscopy (SEM) and energy-dispersive X-ray spectroscopy (EDX) reveal features such as gel morphology, unreacted particles, porosity, and elemental distributions [23]. Studies have shown that dense matrices with sodium aluminosilicate hydrate (N-A-S-H) and calcium aluminosilicate hydrate (C-A-S-H) gels correlate with enhanced strength and reduced permeability [24]. Fourier-transform infrared spectroscopy (FTIR) also provides insights into reaction mechanisms, with characteristic band shifts linked to polycondensation and network formation [25]. By correlating these microstructural features with mechanical and sustainability outcomes, the practicality of SSA-based binders can be better established.

Recent studies have explored diverse activation strategies for low-carbon binders, including nano-silica-enhanced alkali activation, calcined clay activation, and waste-derived silicates. Sodium-based activators remain the most effective due to their high dissolution capacity, whereas potassium-based activators often exhibit rapid viscosity increase and less efficient gel formation [26]. Commercial sodium silicate, while effective, is energy-intensive to produce. Silica fume-derived sodium silicate offers a low-cost, low-emission alternative with high reactive SiO_2 availability. Previous research has shown that nano-silica and natural pozzolans can significantly enhance geopolymerization [27]. While prior studies have synthesized sodium silicate alternatives (SSA) from RHA, SF, and other by-products, most have examined either mechanical performance or chemical reactivity in isolation. Few have provided an integrated evaluation of mechanical properties, microstructural development, and sustainability. In this study, fly ash-based geopolymer mortars were activated with silica fume-derived sodium silicate (SSA) combined with NaOH, and their performance was benchmarked against commercial sodium silicate (SSC). To provide broader context, potassium-based activators (CK and AK) were also included, as KOH is locally used in some regions due to cost, availability, and lower hygroscopicity. Although these mixes underperformed, they offer useful contrasts that inform the selection of activators. The scope of this work includes mechanical testing, fresh property assessment, and sustainability analysis supported by new benchmarking indices for cost (Ic) and CO_2 efficiency (Efc). Microstructural and chemical transformations were examined using SEM, EDX, and ATR-FTIR to establish links between binder chemistry, strength development, and environmental impact. Together, these contributions aim to demonstrate SSA's potential as a sustainable, cost-effective, and technically reliable alternative to SSC for construction applications.

2. Materials

2.1. Silica Fume (SF)

Con-Fume [28]—a commercial reactive microsilica following ASTM C1240 [29] produced by Kryton International in Canada—was utilized as the Silica Fume in the study to make the alternative sodium silicate (SSA) and potassium silicate (PSA). It is composed predominantly of amorphous SiO_2 (>95 wt.%), in gray powder form. The SEM image of raw silica fume (Figure 1) shows the characteristic spherical morphology of the particles, with a smooth surface and size ranging less than 1 μm in diameter. Significant agglomeration is observed due to the high specific surface area (15,000–25,000 m^2/kg) [30,31] and fine particle size, contributing to the geopolymer matrix's densification. According to supplier specifications, silica fume particles have primary spherical sizes in the range of 0.1–0.3 μm, with agglomerated particle sizes typically below 1 μm due to their high surface energy. The reported pozzolanic strength activity index of silica fume typically exceeds 120% at 7 days,

https://doi.org/10.3390/jcs10010022

which is attributed to its amorphous structure, ultra-fine particle size, and high specific surface area.

Figure 1. SEM image of the Silica Fume (SF) at magnification of (**a**) 100 μm, (**b**) 10 μm.

2.2. Fly Ash Class-F

Centralia Class F fly ash [32], in accordance with ASTM C618 [33], manufactured by Lafarge, was used as a Precursor Material (PM) for the geopolymer preparation. According to supplier specifications, this Class F fly ash exhibits a broad particle size distribution ranging from approximately 1 to 100 μm, with a mean particle diameter typically between 15 and 30 μm, and predominantly spherical morphology. Several authors have reported that the fly ash particles are spherical in shape [34,35], as illustrated in Figure 2, which shows SEM images of the fly ash. Class F fly ash consists mainly of an amorphous aluminosilicate glass phase with minor crystalline constituents such as quartz and mullite. The fly ash exhibits adequate pozzolanic activity of 75% at standard testing ages.

Figure 2. SEM images of the Fly Ash (FA) at magnifications of (**a**) 30 μm, (**b**) 10 μm.

Tables 1 and 2 represent the detailed chemical compositions and physical properties of the silica fume and fly ash.

Table 1. Chemical compositions of silica fume and fly ash [28,32].

Chemical Compositions	Silica Fume (SF)	Fly Ash (FA)
Silicon Dioxide (SiO_2)	89–95%	47.1%
Aluminum Oxide (Al_2O_3)	1.0–2.0%	17.40%
Calcium Oxide (CaO)	0.9–1.5%	14.0%
Iron Oxide (Fe_2O_3)	1.5–2.5%	5.7%
Magnesium Oxide	0.5–1.0%	5.40%

https://doi.org/10.3390/jcs10010022

Table 1. *Cont.*

Chemical Compositions	Silica Fume (SF)	Fly Ash (FA)
Total Alkalis as Equivalent Na_2O	0.5–0.9%	N/A [1]
Sulfur Trioxide (SO_3)	0.1–0.5%	0.80%
Loss on Ignition	2.5–5.0%	0.19%
Moisture Content	0.1–0.5%	N/A

[1] Not Provided by Manufacturer.

Table 2. Physical properties of silica fume and fly ash [28,32].

Physical Properties	Silica Fume (SF)	Fly Ash (FA)
Appearance	Light Gray Powder	Tan Powder
Specific Gravity, CSA A3004-13-A4	2.2	2.66
Bulk Density	600–650 kg/m^3	-
Fineness, CSA A3004-13-A3 (45 µm retained)	1–3%	18.30%
Soundness, CSA A3004-13-B5 (Autoclave Expansion)	0.01–0.05%	0.02%
7-day Pozzolanic Strength Activity Index, ASTM C1240-14	120–127%	101%
BET Fineness, ASTM C1240-14 (specific surface)	18–21 m^2/g	N/A [1]
Physical Properties	Silica Fume (SF)	Fly Ash (FA)
Appearance	Light Gray Powder	Tan Powder

[1] Not Provided by Manufacturer.

2.3. Chemicals

Sodium hydroxide (NaOH) pellets from Thermo Fisher Scientific, with a 40 g/mol molecular weight and pH 14, were used to prepare a 10 M NaOH solution. This solution was then used to dissolve silica fume and produce an SSA solution with a SiO_2/Na_2O molar ratio of about 2. Potassium hydroxide (KOH) pellets from Sigma-Aldrich, USA (85% purity, molecular weight 56.11 g/mol, pH 13.5) were used to prepare a 10 M KOH solution. This solution was used to dissolve silica fume and produce a PSA solution with a SiO_2/K_2O molar ratio of 2. The commercial sodium silicate (SSC) was waterglass (Na_2SiO_3), in liquid form, reagent grade, obtained from Sigma-Aldrich in the US. Its composition is 10.6% Na_2O, 26.5% SiO_2, and 62.9% H_2O, with a 1.39 g/mL density. The commercial potassium silicate (PSC) was an anhydrous powder with a molecular weight of 116.11 g/mol from Thermo Fisher Scientific, which was combined with water to produce a solution comprising 53.75% K_2SiO_3 and 46.25% H_2O.

The target silicate modulus (SiO_2/M_2O, M = Na or K) of approximately 2 was calculated based on the reactive SiO_2 content of silica fume and the alkali oxide equivalents derived from hydroxide solutions. The silica fume used contains approximately 95 wt.% SiO_2. A 10 M NaOH solution provides 10 mol of NaOH per liter, where two moles of NaOH are stoichiometrically equivalent to one mole of Na_2O; thus, 10 mol NaOH corresponds to 5 mol Na_2O. To achieve a SiO_2/Na_2O molar ratio of 2, 10 mol of SiO_2 is required, corresponding to approximately 600 g of SiO_2. Accounting for silica fume purity, this equates to about 630 g of silica fume per liter of 10 M NaOH solution. The same approach was applied for potassium-based activators, where 10 M KOH provides 5 mol of K_2O equivalents, requiring an identical SiO_2 input to maintain a SiO_2/K_2O ratio of ~2.

https://doi.org/10.3390/jcs10010022

3. Methodology

3.1. Mix Design

Preliminary trials at AA/P ratios of 0.4 and 0.8 showed poor workability: the 0.4 mix was too stiff to mold, while the 0.8 mix exhibited flash setting. Therefore, ratios of 0.5 and 0.7 were selected as practical bounds [36], allowing comparison between low- and high-alkali conditions. A Sand/P ratio of 2:1 was fixed, following prior studies, to ensure consistent workability and strength. Table 3 presents the mix proportions for geopolymers prepared with commercial sodium silicate (CS), alternative sodium silicate (AS), commercial potassium silicate (CK), and alternative potassium silicate (AK).

Table 3. Mix compositions of geopolymer mortars.

Mix Code	FA (kg)	Total AA (kg)	Alkaline Activators (AA)						Sand (kg)
			SSA (kg)	SSC (kg)	NaOH 10 M (kg)	PSA (kg)	PSC (kg)	KOH 10 M (kg)	
CS0.7	600	420		214.63	205.37				1200
AS0.7		420	228.71		191.31				
CS0.5		300		153.3	146.7				
AS0.5		300	163.4		136.6				
CK0.5		300					105.53	194.47	
AK0.5		300				162.77		137.24	

The AA used in these mixes is a binary solution of either sodium or potassium silicate (Na_2SiO_3 or K_2SiO_3) and sodium or potassium hydroxide (NaOH or KOH). The volumetric proportion Na_2SiO_3/NaOH and K2SiO3/KOH is maintained at 1:1, and the densities of NaOH, SSC, SSA, KOH, PSC, and PSA are 1.33, 1.39, 1.59, 1.29, 0.7 and 1.53 g/mL, respectively. Table 4 shows the volumetric proportions of the mixes.

Table 4. Volumetric proportions of geopolymer mortars.

Mix code	Sand/P	AA/P Ratio	M_2SiO_3/MOH *	MSC (L)	MSA (L)	MOH 10 M (L)
CS0.7	2:1	0.7	1:1	154.41		154.41
AS0.7					143.84	143.84
CS0.5		0.5		110.3		110.3
AS0.5					102.74	102.74
CK0.5				150.75		150.75
AK0.5					106.38	106.38

*: M = Na or K.

3.2. Preparation of Alkali Activator (AA)

The NaOH solution was prepared by dissolving 400 g of sodium hydroxide pellets per 1000 mL of water to obtain a solution with the desired molarity of 10 M. The KOH solution was prepared by dissolving 561.1 g of potassium hydroxide pellets per 1000 mL of deionized water to obtain a solution with the desired molarity of 10 M. In contrast, the SSA and PSA solutions were designed using Equations (1) and (2) and prepared by dissolving 632 g of silica fume per 1000 mL of NaOH or KOH solution with a molarity of 10 M [20]. The mixture was stirred for 30 min and left to react for 24 h in a closed container to ensure thorough dissolution of the silica fume in the hydroxide solution before subsequent use. Furthermore, a SiO_2/M2O molar ratio of 2 (M = Na or K) and the purity of SF used in the current study (95% SiO_2 with a molecular weight of 60 g/mol) were considered.

$$2SiO_2 + 2NaOH \rightarrow Na_2O\ (SiO_2) + H_2O \quad (1)$$

https://doi.org/10.3390/jcs10010022

$$2SiO_2 + 2KOH \rightarrow K_2O(SiO_2) + H_2O \quad (2)$$

3.3. Preparation of Geopolymer Mortar Specimens, Curing and Testing Methods

The fly ash and sand were accurately weighed as per the proportions listed in Table 3 and mixed in a table mixer for 2 min until a uniform dry mixture was achieved. The pre-mixed binary alkali activator solution (SSC or SSA and NaOH, or PSC or PSA and KOH) was slowly added to the dry mixture and mixed for two more minutes. After thorough mixing, the flowability of the fresh mortar was assessed using the flow table test as per ASTM C1437 [37] at an ambient temperature of 23 ± 2 °C. The time of setting was determined using the Vicat apparatus according to ASTM C807 [38]. Additionally, specimens in the form of 50 × 50 × 50 mm cubes, dog-bone (briquette) specimens, and 30 × 30 × 100 mm prisms were prepared following ASTM standards C109 [39], C307 [40], and C293 [41], respectively, for compressive, tensile, and flexural strength tests. Due to limitations in standards specific to geopolymer, relevant and comparable mortar testing methods have been used.

The prepared specimens were left to cure for 24 h in the molds at ambient temperature. After demolding, the hardened mortar samples underwent heat curing in an oven for 72 h. After 72 h heat curing, specimens were allowed to cool gradually inside the oven to room temperature to prevent thermal shock and microcrack formation. Subsequently, the specimens were kept at ambient temperature until the day of testing. This curing regime was chosen because geopolymers typically experience rapid reactions during the early stages of curing, especially within the first few days [42]. Geopolymer curing temperatures and heat curing times can vary significantly, ranging from 40 °C to 800 °C and from 0 to 28 days, respectively. However, elevated temperatures are often required to accelerate the geopolymerization process and achieve sufficient strength development within a short time frame [43]. Higher curing temperatures contribute to increased CO_2 emissions, which detract from the sustainability of the process. For this study, two batches of AS0.7 were cured at 65 °C and 80 °C (AS0.7-65, AS0.7-80), while one batch of CS0.7 was cured at 80 °C (CS0.7-80). Since the effect of the higher temperature was minimal on the strength development, the rest of the batches were cured at 65 °C. The reported results are the average of three specimens tested on 7, 14, and 28 days.

3.4. SEM Imaging and EDS Analysis

SEM and EDS analyses were performed on sodium-based geopolymer formulations to investigate the formation and composition of geopolymerization products and compare the effect of sodium silicate and SSA activators. Rectangular specimens of approximately 5 × 5 × 2.5 mm were cut from hardened beam samples. The specimens were polished using a ToronPol polishing machine, from Torontech Inc., Toronto, ON, Canada, equipped with abrasive paper at 400 rpms for 3 min to obtain a surface flat enough for microscopy. The specimens were vacuum-dried for one week in a Hitachi ZoneSEM desktop cleaner (Hitachi High-Tech Corporation, Tokyo, Japan), which removes hydrocarbon contamination from specimens using UV light. The specimens were taken out of the vacuum and quickly coated with carbon using a Cressington 208 carbon high vacuum carbon coater (Cressington Scientific Instruments Ltd., Watford, UK) to ensure the preservation of the microstructure and conductivity for electron microscopy. A Scanning Electron Microscope (SEM) was used at a 1–15 kV voltage on thin polished sections using a Hitachi S-4800 field emission SEM (Hitachi High-Tech Corporation, Tokyo, Japan). SEM images were used to recognize the reaction of geopolymer precursor material with solutions in the internal microstructure and identify entities such as amorphous gel and crystalline phases, unreacted or partially reacted particles, microcracks and pores, and Interfacial Transition Zone (ITZ) of aggregates and geopolymer paste matrix. The Hitachi S-4800 is equipped with a Bruker Quantax Energy Dispersive Spectroscopy (EDS) system (Bruker Nano GmbH, Berlin, Germany), which was used at an acceleration voltage of 8–13 kV and a working distance of 15 mm to optimize X-ray collection and detect the elemental composition of geopolymerization products. Quantax software (version 7.3) was used to generate high-resolution elemental maps while the Hyperspy (version 2.2, https://pypi.org/project/hyperspy/2.2.0/ accessed on 20 January 2025) [44] python package was used to generate X-ray spectra graphs and quantify elemental compositions. Comparative imaging and mapping were performed on different mixes to identify variations in morphology and validate the formation of N-A-S-H and C-A-S-H gels.

3.5. ATR-FTIR Analysis

The ATR-FTIR analysis was conducted to investigate the structural and chemical changes occurring in the silicate components of silica fume, fly ash, and geopolymer formulations due to alkali activators. This analysis aimed to identify characteristic functional groups, monitor shifts in vibrational bands, and assess the geopolymerization process. The samples were manually grounded into a fine powder in a mortar pestle and placed directly over the ZnSe ATR crystal of the Agilent Cary 630 FTIR spectrometer. The samples were analyzed in the spectral range of 4000 to 400 cm^{-1} at room temperature with 32 scans and 4 cm^{-1} resolutions and support from the Agilent MicroLab software (version 5.4). The analysis specifically targeted changes in the Si-O and Al-O bonding environments, using ATR-FTIR spectra to identify peak shifts in silica fume and fly ash after alkali activation. These shifts indicate the dissolution of raw materials and the formation of new silicate and aluminosilicate structures, such as Si-O-Al linkages, which are essential for geopolymer formation. Additionally, the disappearance of peaks related to hydrates confirmed the completion of the curing process.

4. Results and Discussion

4.1. Fresh Properties

4.1.1. Time of Setting

Fresh mortars were introduced into the Vicat apparatus mold within five minutes of initial contact between the alkali activator and the precursor material. As per ASTM C807, the time of setting is defined as the point at which the Vicat needle penetrates to a depth of 10 mm (Figure 3). Findings indicate that sodium-based activators (SSA and SSC) exhibit longer setting times, whereas potassium-based activators (PSA and PSC) set more rapidly. This variation is influenced by the alkaline activator to precursor (AA/P) ratio, which affects the dissolution rate of aluminosilicate precursors. A higher AA/P ratio increases the availability of reactive ions (OH^-, SiO_4^{4-}, Na^+, and $Al(OH)_4^-$), accelerating polymerization and gel formation. Notably, SSC-based mixes set faster than SSA-based ones due to the higher reactivity of pre-dissolved SiO_4^{4-} ions in SSC, which immediately polymerize with $Al(OH)_4^-$ and Na^+. In contrast, SSA requires gradual dissolution of silica fume in NaOH, delaying the availability of reactive silicates despite its higher OH^- content. Additionally, potassium-based activators (PSA and PSC) exhibit the fastest setting times due to the larger ionic radius [45] and higher mobility of K^+ ions, which enhances charge balancing and condensation reactions [46], leading to rapid gel network formation. Potassium-based activators showed lower mechanical performance primarily due to their chemical behavior during geopolymerization. The larger ionic radius of K^+ reduces its ability to stabilize aluminosilicate gel structures compared to Na^+, leading to weaker N-A-S-H network formation. Additionally, potassium silicate dissolves more slowly, limiting the early availability of reactive silicate species. This slower dissolution causes early viscosity

https://doi.org/10.3390/jcs10010022

buildup and reduced workability, resulting in less dense matrices and lower strength in CK and AK mixes.

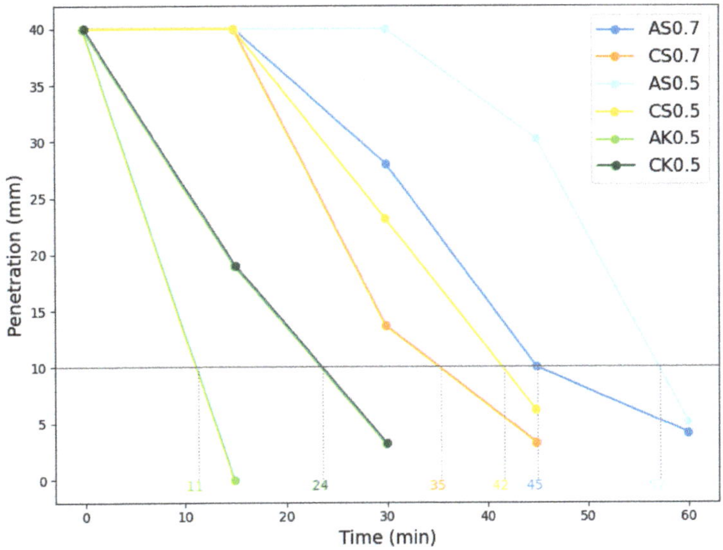

Figure 3. Penetration vs. time: Time of setting corresponds to the moment when the needle penetrates 10 mm in the mortar.

4.1.2. Flowability

Table 5 presents the flow percentage of each mix, obtained using the jump table according to ASTM C1437. The test involved filling the cone with mortar, lifting it, and dropping the table 25 times within 15 s. Figure 4 depicts the flow table test. The spread diameter of the mortar was then measured using the formula below, where A is the average flow diameter of four readings in mm and D_{org} is the original base diameter (100 mm).

$$Flow\ (\%) = \left(\frac{A - D_{org}}{D_{org}}\right) \times 100 \quad (3)$$

Table 5. Flowability test.

	AS0.7	CS0.7	AS0.5	CS0.5	CK0.5	AK0.5
Flow Diameter (A) (mm)	250	245	240	250	235	150
Flow (%)	150	145	140	150	135	50

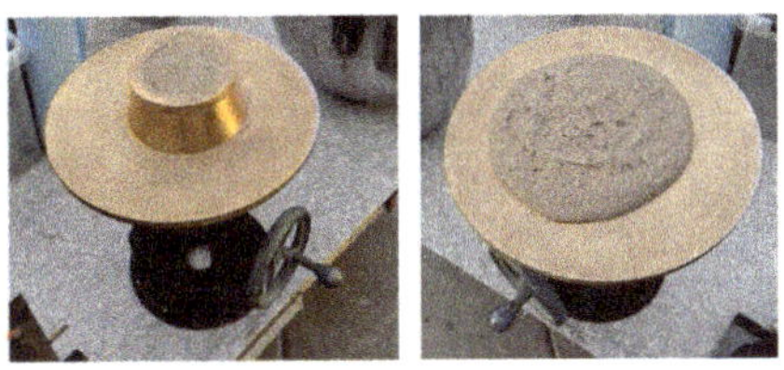

Figure 4. Flow table test.

https://doi.org/10.3390/jcs10010022

Results show that sodium-based activators (SSA and SSC) provide higher flowability than potassium-based activators (PSA and PSC). CK0.5 exhibits the lowest flowability (50%), highlighting the reduced workability of K-based geopolymers. The lower flowability of K-based mixes correlates with their faster setting times, as higher K^+ ions accelerate polycondensation, causing early viscosity buildup and stiffening, reducing flow. Conversely, sodium-based activators—particularly SSA—react more gradually, improving dispersion and fluidity. Additionally, the AA/P ratio influences flowability, with higher ratios (0.7) improving fluidity. However, even at the same AA/P ratio, K-based activators consistently lower flowability, likely due to their faster setting behavior. Thus, sodium-based activators enhance workability, while K-based mixes may require water adjustments or plasticizers to maintain optimal flow without affecting setting time. The use of superplasticizers was not investigated in this study due to uncertainties regarding their behavior and long-term compatibility in highly alkaline geopolymer matrices. Although previous studies have reported that conventional superplasticizers can function in geopolymer systems, their efficiency and stability often differ from those in OPC-based mixtures because of the high alkalinity and distinct reaction mechanisms. Therefore, the compatibility and performance of admixtures in geopolymer mortars are identified as an important area for future research.

4.2. Hardened Properties

4.2.1. Compressive Strength (f'_c)

In this study, a Forney compression machine with a maximum capacity of 3000 KN was used, and an applied loading rate was maintained at 0.15 MPa/s for every test. The compressive strength is calculated as follows:

$$f'_c = \frac{P}{A} \tag{4}$$

where f'_c is the mortar's compressive strength in MPa, P is the total maximum load in N, and A is the area of loaded surface mm^2.

The results in Figure 5 show that CS0.7-80 exhibited compressive strengths 20–35% lower than AS0.7-80 and AS0.7-65 at all ages. Unexpectedly, AS0.7-65 outperformed AS0.7-80, contrary to typical literature trends. This anomaly is attributed to rapid geopolymerization at 80 °C, which may produce a rigid but less uniform gel network and induce thermal microcracking or shrinkage, thereby limiting further reaction of fly ash particles. SEM analysis, discussed in Section 4.4.1, supports this interpretation, showing more unreacted fly ash and microcrack features in AS0.7-80, while AS0.7-65 developed a denser, more homogeneous matrix that enabled continued geopolymerization.

At AA/P = 0.5, both CS0.5 and AS0.5 reached ~40 MPa at 28 days, though AS0.5 showed higher strengths at 7 and 14 days. This agrees with reports that reducing AA/P generally improves CS-based systems, but in AS mixes, the higher content of fine SF particles at AA/P = 0.7 promoted superior strength gain relative to AS0.5.

Strength development was largely completed within the first 7 days, with little or no gain thereafter, consistent with Davidovits [42]. he 28-day strengths of SSA mixes in this study (39–45 MPa) compare well with literature: ~43 MPa for fly ash binders with SF-derived activators [19], ~50 MPa for GGBS binders with RHA-derived activators [22], and ~50 MPa for GGBS concrete with SF-derived activators [22]. This benchmarking demonstrates that SSA-based mixes not only outperform SSC-based mortars but also achieve strengths competitive with other waste-derived activator systems. Quantitative porosity measurements were not conducted in this study; therefore, interpretations regarding porosity and microcrack density are based on qualitative SEM observations and are acknowledged as a limitation of the present work. Although geopolymers are often

https://doi.org/10.3390/jcs10010022

reported to exhibit high early-age strength, this behavior depends strongly on precursor type, activator chemistry, and curing conditions. In this study, the fly ash-dominant system and controlled alkali availability led to slower dissolution and progressive geopolymerization, resulting in a strength-gain trend similar to conventional concrete. Such behavior is consistent with Class F fly ash–based geopolymers reported under comparable curing regimes. The slight reduction in 28-day compressive strength is attributed to moisture loss and microstructural stabilization following early geopolymerization, which may induce minor microcracking during curing. Similar behavior has been reported for heat-cured fly ash–based geopolymers. Extended curing periods (e.g., 90 days) are expected to result in strength stabilization rather than continued decline; however, long-term mechanical testing and durability studies are required to confirm this trend and are recommended for future work.

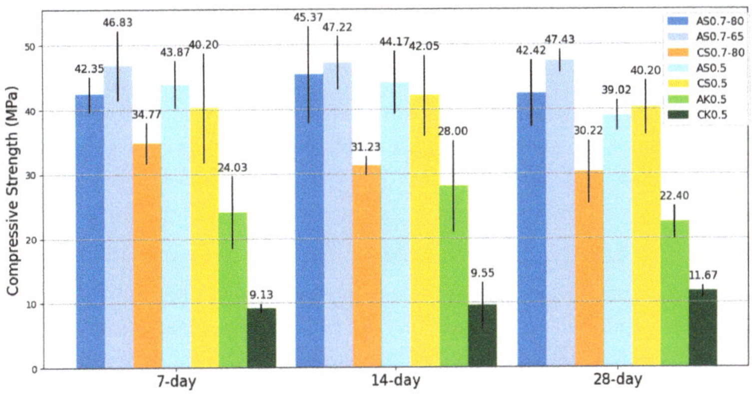

Figure 5. Compressive strength (MPa) at 7, 14, and 28 days.

4.2.2. Flexural Strength (Fr)

The Pasco mini MTM, capable of measuring up to 7100 N (newtons) of force, was used to perform the three-point bending tests on the 30 × 30 × 100 prisms and measure the peak load (P). The modulus of rupture is calculated as:

$$F_r = \frac{3PL}{2bd^2} \tag{5}$$

where F_r is the modulus of rupture in MPa, and P is the maximum applied load in N. L, b, and d are the span length, width, and depth of the prism (and equal to 88, 30, and 30 mm), respectively. The support's distance from the beam's edges was kept at 6 mm.

Figure 6 shows that the mixes AS0.7-80 and AS0.7-65 demonstrated higher Modulus of Rupture (MOR) values of 3.56 MPa and 2.90 MPa, respectively, at 28 days compared to CS0.7-80, which showed an MOR of 1.42 MPa. These results represent increases of approximately 150.7% for AS0.7-80 and 104.2% for AS0.7-65 over CS0.7-80. However, CS0.5 exhibited a superior MOR of approximately 5 MPa across all three testing ages, surpassing all other mixes. Flexural strength development showed minimal dependence on time, with an almost flat trendline for all mixes.

https://doi.org/10.3390/jcs10010022

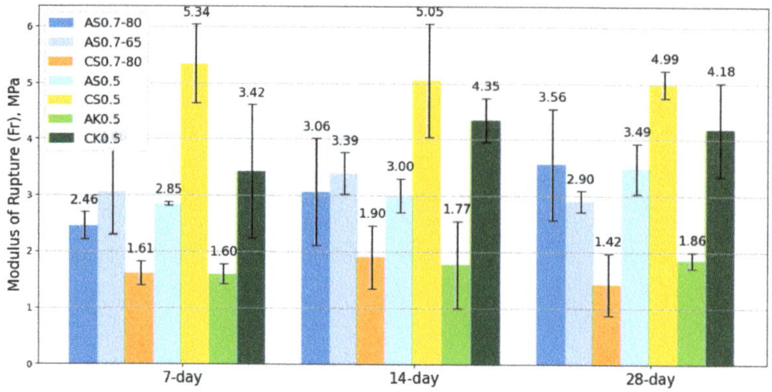

Figure 6. Flexural strength (MPa) at 7, 14, and 28 days.

4.2.3. Tensile Strength (Ft)

The Pasco mini MTM was used to perform direct tensile tests on the dog-bone (briquet) specimen and measure the peak load (P). The tensile strength (F_t) is equal to the stress calculated at maximum load. It is calculated as below, where b and d both equal 25.4 mm.

$$F_t = \frac{P}{bd} \quad (6)$$

As shown in Figure 7, specimens with an AA/P ratio of 0.5 excel in tensile tests compared to those with an AA/P ratio of 0.7, achieving tensile strengths of 1 MPa for CS0.5 and 0.86 MPa for AS0.5 at 28 days. Additionally, AS0.7-80 and AS0.7-65 outperform CS0.7-80 in tensile tests, showing 76% and 45% higher strength at 28 days, respectively.

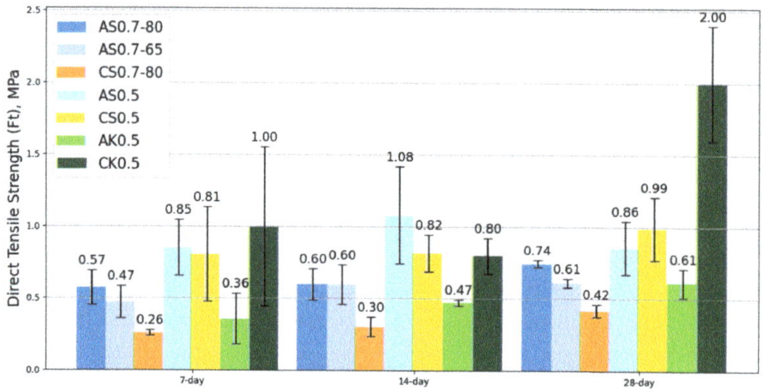

Figure 7. Tensile strength (MPa) at 7, 14, and 28 days.

The compressive strength (f'_c), flexural strength (F_r), and tensile strength (F_t) results illustrate that CS0.5 exhibits robust mechanical performance, with strengths reaching 40 MPa, 5 MPa, and 1 MPa, respectively, by 28 days. AS0.5 displayed comparable performance with values of 39 MPa for compressive strength, 3.49 MPa for flexural strength, and 0.86 MPa for tensile strength. making it suitable for applications requiring structural integrity and strength. The lower AA/P ratio (0.5) is key in enhancing the material's overall strength, as the denser microstructure provides better resistance to compressive, bending, and tensile forces. Mixes like AS0.7-80 and AS0.7-65 perform better than CS0.7-80, especially in flexural and tensile strengths, and their compressive strengths are higher than all other

https://doi.org/10.3390/jcs10010022

mixes. This suggests that the alternative sodium silicate and lower curing temperatures (65 °C) in AS mixes contribute to improved bending and tension performance. However, potassium-based mixes (CK0.5 and AK0.5) show significantly lower results in all categories, indicating sodium-based activators are more effective for achieving high strength. Overall, CS0.5 and AS0.5 stand out as the better-performing mixes in all mechanical tests, while AS0.7-65 and AS0.7-80 offer good performance in compression.

The trends observed in flexural and tensile strength are consistent with the compressive strength behavior and can be attributed to microstructural stabilization and the evolution of interfacial bonding within the geopolymer matrix. Minor microcracking induced by moisture loss and thermal effects during curing is more critical under tensile and flexural loading, leading to slightly reduced strengths at later ages.

4.3. Sustainability

4.3.1. Cost

In this section, we evaluate the cost-effectiveness of CS, AS, CK, and AK by comparing the expenses associated with their respective alkali activator (AA) sources and sodium/potassium silicate options. The varying parameters between the mixes are the choice of silicate source (SSC/SSA or PSC/PSA) and the proportion of alkali activator. The commercial sodium silicate (SSC) was procured at 78 CAD per 1 L. The cost to produce 1 L of NaOH 10 M was determined to be 45.76 CAD (derived from 400 gr NaOH pellets + 1000 mL H_2O) (Sigma-Aldrich [47]). Silica fume was priced at an average of CAD 0.5 per kg [48]. The commercial potassium silicate (PSC) was purchased at 363 CAD per 2 kg. The cost to produce 1 L of KOH 10 M was determined to be 53.42 CAD (derived from 561.1 gr KOH pellets + 1000 mL H_2O) (Sigma-Aldrich, St. Louis, MO, USA [49], Thermo Scientific Chemicals, Waltham, MA, USA [50]). Based on the calculations, the SSA + NaOH alkali activator demonstrates higher cost-effectiveness, with a cost of 74.15% compared to the SSC + NaOH (Figure 8a). Notably, K-based activators are not only higher in cost but also less effective in the strength gain of the geopolymer, making their use impractical. Figure 8a demonstrates the cost of alkaline activators in CAD per liter. The cost analysis is limited to a material-based comparison, considering only the prices of alkaline activators and their constituent raw materials. Labor, energy, and processing costs associated with activator preparation were not included and are acknowledged as limitations of this study.

4.3.2. CO_2 Emissions

Carbon footprint data of individual ingredients were obtained from published sources and are summarized in Figure 8. The CO_2 emissions considered were: sodium silicate (1.22 kg CO_2 e/kg [51]), NaOH (1.22 kg CO_2 e/kg [52]), KOH (0.77 kg CO_2 e/kg [53]), and silica fume (0.014 kg CO_2 e/kg [54]). Potassium silicate was estimated at ~2.5 kg CO_2 e/kg, based on reported contributions from silica production, potassium carbonate synthesis, manufacturing, and transport. These values represent material production only; transportation and the energy required for elevated curing were not considered, which is acknowledged as a limitation.

4.3.3. Compressive Strength Cost Index (Ic)

The compressive strength cost index Ic is calculated to evaluate the economic efficiency of geopolymer materials [55]. This index, measured in CAD/m^3/MPa, is given by the formula:

$$I_C = \frac{C_t}{f'_{c28}} \tag{7}$$

https://doi.org/10.3390/jcs10010022

where Ic represents the compressive strength cost index, C_t (CAD/ m^3) is the cost of geopolymer per cubic meter, and f'_{c28} (MPa) represents the 28-day compressive strength of the geopolymer. A smaller compressive strength cost index indicates a relatively lower overall cost of using geopolymer materials. The costs of Fly ash and sand were averaged to be 32 CAD [56] and 48 CAD [57] per ton, respectively. Geopolymer mixes with alternative sodium silicate (AS0.7 and AS0.5) show superior performance with an Ic of approximately 30, indicating better economic efficiency. In contrast, the CS0.7 mix has an Ic of about 66.86, and the K-based mixes have indices of 64.22 and 73.86, indicating that they are over twice as costly per unit of strength. The CS0.5 mix, with an Ic of 40, demonstrates significant strength at 28 days but remains less cost-effective than the AS0.5 and AS0.7 mixes (Table 6).

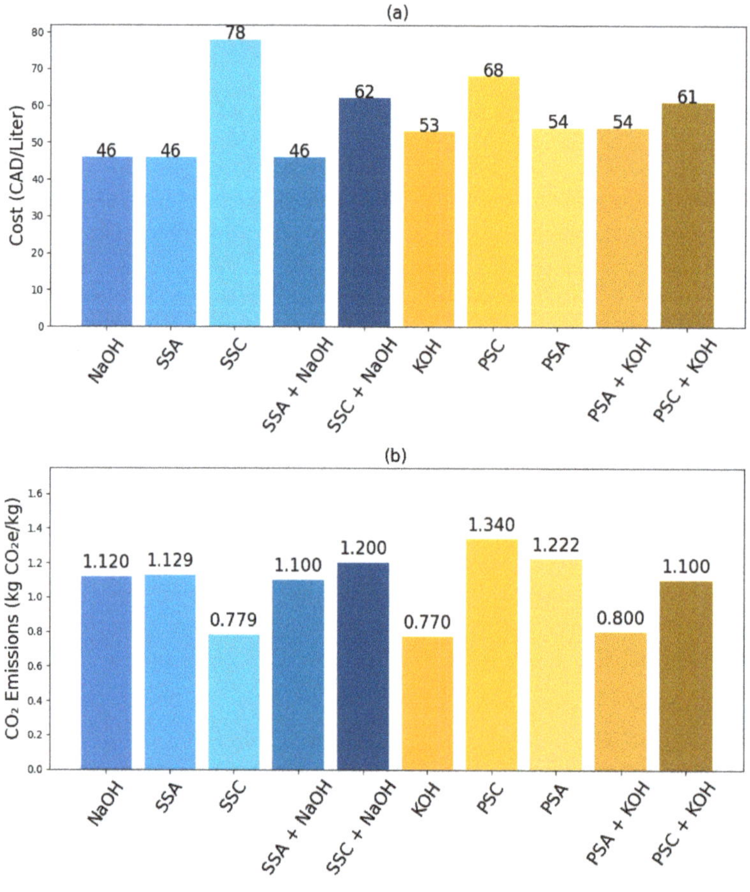

Figure 8. (**a**) Cost (**b**) CO_2 emissions of single and binary alkali activators per liter.

Table 6. CO_2 emission and sustainability index of geopolymer mixes.

	AS0.7	CS0.7	AS0.5	CS0.5	CK0.5	AK0.5
28-day f'_c (MPa)	44.77	30.22	39.02	40.20	11.67	22.4
Production Cost (CAD/m^3)	1388.03	1931.89	1104.08	1533.49	1398.74	1494.60
CO_2-m^3 (kg CO_2)	505.65	509.75	399.61	408.26	258.42	284.17
Compressive cost index (I_c)	30.99	63.93	28.30	38.15	119.80	66.74
Sustainability index E_{fc}	10.66	16.86	10.24	10.15	10.21	12.96

https://doi.org/10.3390/jcs10010022

4.3.4. Sustainability Index (Efc)

To evaluate the sustainability of construction materials, we can calculate the Sustainability Index (E_{f_c}) [55] as shown in Equation (7). This index normalizes the CO_2 emissions of concrete mixes according to their 28-day compressive strength. This method provides a more comprehensive view of the environmental impact by considering both the CO_2 emissions and the material's performance.

$$E_{f_c} = \frac{CO_{2-m^3}}{f'_{c28}} \quad (8)$$

The CO_2 emissions of Fly ash and sand were taken as 0.04 [58] and 0.0077 [59] kg CO_2 e/kg, respectively. A smaller Sustainability Index indicates a lower overall CO2 emission from geopolymer materials. The analysis of the Sustainability Index (E_{f_c}) reveals that the CS0.5 mix is the most sustainable, with the lowest E_{f_c} of 10.15, representing the best balance between low CO_2 emissions and high compressive strength. The AS0.5 mix follows closely with an E_{f_c} of 10.24, also demonstrating high sustainability. The AS0.7 mix, though having higher emissions, maintains a relatively low E_{f_c} of 10.66, making it a good sustainable choice. In contrast, the CS0.7 mix has the highest E_{f_c} of 16.86, indicating higher CO2 emissions per unit of compressive strength. Potassium-based mixes, CK0.5 and AK0.5, have E_{f_c} values of 10.21 and 12.96, respectively, reflecting more significant environmental impact compared to sodium-based mixes. Overall, the AS0.5 and CS0.5 mixes are highlighted as the most eco-friendly options, effectively balancing material efficiency and carbon footprint (Table 6).

Reported E_{f_c} values for ordinary Portland cement (OPC) concretes typically range from 15 to 20, based on emissions of ~300–400 kg CO_2/m^3 and 28-day strengths of 20–30 MPa [60]. Geopolymer concretes activated with commercial sodium silicate often report values around 13–18, reflecting improved but still significant embodied CO_2 [61]. In comparison, the SSA mixes in this study demonstrate a clear sustainability advantage, outperforming both OPC and conventional geopolymers.

4.4. Microstructural Analysis

4.4.1. SEM

In SEM imaging of fly ash-based geopolymers, the amorphous and crystalline phases exhibit distinct morphologies that reflect the geopolymerization process and variations in material composition. When sodium silicate is used as the activator, the primary reaction product is an amorphous sodium aluminosilicate hydrate (N-A-S-H) gel. This amorphous phase typically forms a continuous, dense, and homogeneous matrix [62], embedding any crystalline phases present [63], as shown in Figure 9 of CS0.7, where the smooth gel phase covers the imaged surface with few distinct entities and morphologies. The matrix appears smooth or slightly granular, with minimal defined boundaries between particles, indicating efficient geopolymerization [64,65]. In contrast, crystalline phases display discrete, well-defined morphologies, including needle-like, flocculent, or plate-like structures [66], as observed in Figure 10b. These crystalline features typically form on the surfaces of partially or fully reacted fly ash particles and are associated with specific reaction products, including calcium silicate hydrate (C-S-H), calcium hydroxide ($Ca(OH)_2$), and minerals such as mullite and zeolite [67]. The spherical voids in Figure 10b are remnants of fly ash particles attacked by the alkali and merged into the gel paste. The transition from amorphous to crystalline phases has been associated with lower strengths in geopolymer systems [68]. A qualitative comparison of the interfacial transition zone (ITZ) indicates that the commercial sodium silicate (CS) mortar exhibits a wider and less homogeneous ITZ with visible microcracks and partially reacted fly ash near the aggregate surface. In

https://doi.org/10.3390/jcs10010022

contrast, the silica fume-derived sodium silicate (AS) mortar shows a denser and more continuous ITZ, with the aggregate well embedded in a compact N(C)-A-S-H gel matrix (Figure 10). Quantitative ITZ measurements were not conducted and is acknowledged as a limitation; therefore, this assessment is based on qualitative SEM observations.

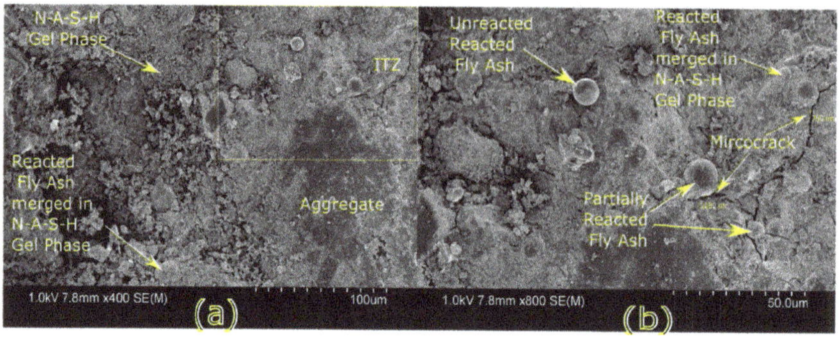

Figure 9. SEM Images of Specimen CS0.7 at magnification of (**a**) 100 μm, (**b**) 50 μm.

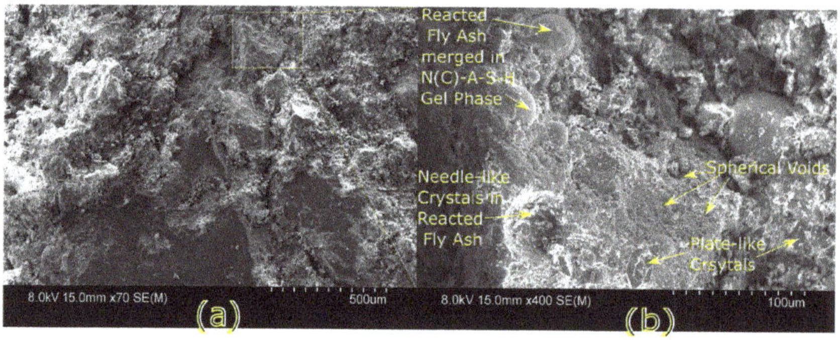

Figure 10. SEM Images of Specimen AS0.7-80 at magnification of (**a**) 500 μm, (**b**) 100 μm.

The choice of alkaline activator significantly impacts the phase formation and microstructure of fly ash-based geopolymers [24]. Hydroxide-based activators, such as sodium hydroxide (NaOH), facilitate the dissolution of aluminosilicate precursors, increasing the availability of reactive species necessary for crystallization. This results in the formation of well-defined crystalline phases. On the other hand, silicate-based activators, like sodium silicate (Na_2SiO_3), promote the generation of an amorphous gel phase, leading to a more cohesive and homogeneous matrix. The combination of NaOH with silica fume produces results comparable to, and in some cases superior to, commercial sodium silicate in terms of matrix density and mechanical properties. Loose partially reacted or unreacted fly ash particles are also evident in SEM images, appearing as spherical entities embedded within the matrix. Their presence signifies incomplete geopolymerization, which correlates with higher porosity and lower mechanical strength (Figure 11). These particles are more prevalent in CS mixes compared to AS (Figure 12), which is also reflected in the significantly lower compressive, tensile, and flexural strengths of CS0.7 compared to AS0.7-80 and A0.7-65 as described in Section 4.2.

https://doi.org/10.3390/jcs10010022

Figure 11. SEM Images of Specimen CS0.7 at magnification of (**a**) 100 µm, (**b**) 30 µm.

Figure 12. SEM Images of Specimen AS0.7-80 at magnification of (**a**) 100 µm, (**b**) 30 µm.

Figures 13 and 14 reveal a continuous mass of a sodium (calcium) aluminosilicate N(C-)A-S-H matrix and partially reacted fly ash particles. Various products with distinct morphologies have formed on the surface of the fly ash particles. Regular-shaped products of calcium silicate hydrate (CSH) and aluminum-silicate (Al-Si) gel are observed on these particles, indicating the development of a dense geopolymer matrix. The strength development of AS0.7 and AS0.5 can be attributed to the presence of microsilica particles, densifying the porous microstructure of the geopolymer mix. The SEM images show fine silica fume particles in AS0.5 and AS0.7 mixtures, contrasting with CS0.7 and CS0.5, which exhibit a more porous microstructure. This suggests that the incorporation of silica fume contributes to the enhancement of the microstructure by reducing voids and refining the matrix.

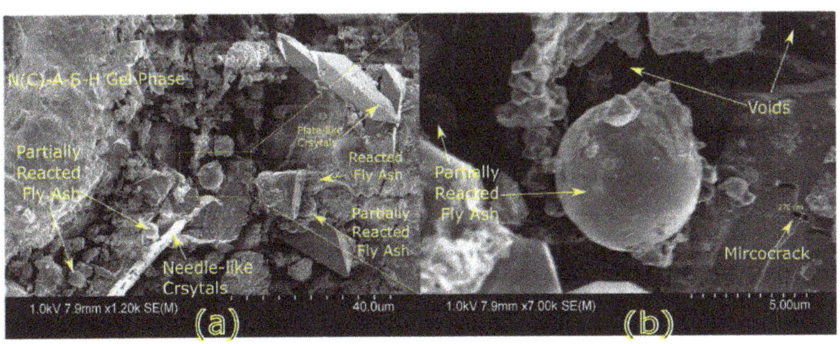

Figure 13. SEM Images of Specimen AS0.7-65 at magnification of (**a**) 40 µm, (**b**) 5 µm.

https://doi.org/10.3390/jcs10010022

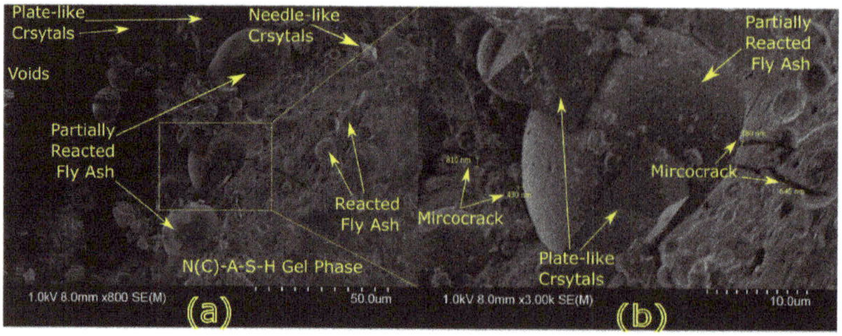

Figure 14. SEM Images of Specimen AS0.7-65 at magnification of (**a**) 50 μm, (**b**) 10 μm.

The formation of sodium aluminosilicate hydrate, the primary reaction product of geopolymerization, results in an amorphous gel matrix. This matrix forms through interactions between fly ash glassy spheres and the alkaline solution, signifying effective geopolymerization and the transition from a loose fly ash structure to a cohesive network. Alkali activators adhere to fly ash surfaces, filling pores and intertwining to create a robust network. A significant amount of unreacted fly ash remains in CS0.5, exhibiting larger pores and a looser structure compared to AS0.5 (Figures 15 and 16). Fewer partially reacted spheres are encapsulated in gel products in CS0.5, while AS0.5 demonstrates a more consolidated microstructure.

Figure 15. SEM Images of Specimen AS0.5 at magnification of (**a**) 50 μm, (**b**) 20 μm.

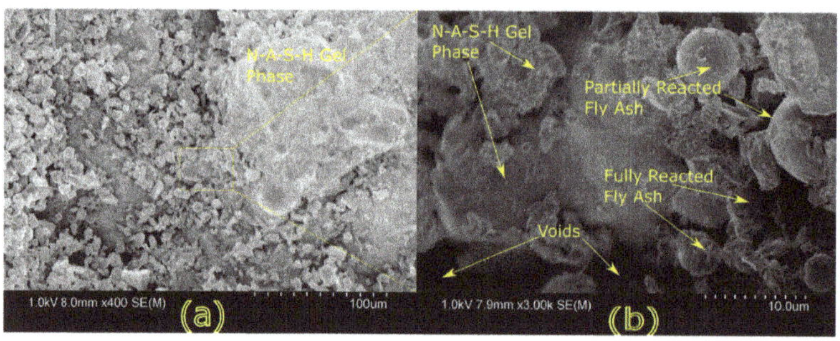

Figure 16. SEM Images of Specimen CS0.5 at magnification of (**a**) 100 μm, (**b**) 10 μm.

4.4.2. EDX

The EDX map and spectra for Specimen CS0.7, shown in Figure 17, provide valuable insights into the elemental composition and distribution of the geopolymer matrix. The matrix is predominantly composed of oxygen (37.49 wt.%), silicon (27.13 wt.%), and sodium

https://doi.org/10.3390/jcs10010022

(16.70 wt.%), consistent with a sodium aluminosilicate hydrate (N-A-S-H) gel structure. Aluminum contributes 10.59 wt.%, reflecting its integral role in forming aluminosilicate linkages, while magnesium is present at 4.34 wt.%. The relatively low calcium content (3.76 wt.%) suggests limited reactivity, likely due to incomplete geopolymerization of the fly ash, which may be attributed to insufficient availability of OH^- ions and an excess of silicate ions (SiO_4^{4-}) in CS0.7 with sodium silicate as the activator.

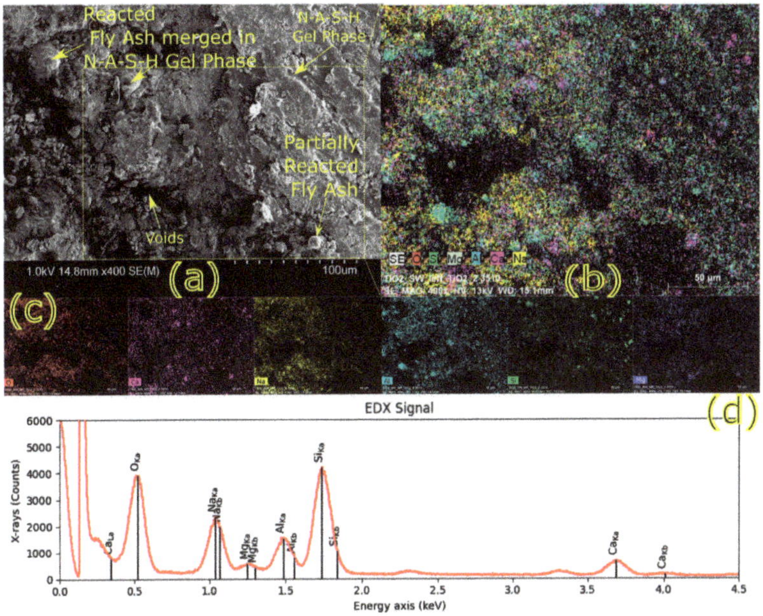

Figure 17. (**a**) SEM Image of CS0.7 at magnification of 100 µm, (**b**) EDX map, (**c**) Elemental maps, (**d**) EDX Spectra.

This lack of calcium incorporation aligns with the observed amorphous microstructure of CS mixes. Figure 17d shows characteristic peaks in the x-ray (counts) vs. energy (keV) graph, where calcium shows distinct excitation energies of $K\alpha$ = 3.75 keV, $K\beta$ = 4.00 keV, and $L\alpha$ = 0.3 keV, further corroborating its minor presence in the matrix.

The EDX map for specimen AS0.7-80 (Figure 18e) captures the microstructural and elemental characteristics near the interfacial transition zone (ITZ) of geopolymer paste and aggregate. The presence of crystalline phases enriched with calcium is particularly in the plate-like structures. Silicon and sodium remain dominant in the overall composition (Figure 18c), while the elevated calcium concentration in crystalline regions supports the formation of calcium hydroxide ($Ca(OH)_2$) crystals and calcium aluminosilicate hydrate (C-A-S-H) gel and semi-crystalline phases. Calcium in fly ash reacts preferentially with OH^- ions, forming calcium hydroxide ($Ca(OH)_2$), which precipitates rapidly in an alkaline environment [69]. This reaction occurs faster than calcium's interaction with silicate or aluminosilicate species. However, when sufficient silicate is available, Ca^{2+} can also contribute to the formation of calcium silicate hydrate (C-S-H), and the coexistence of C-S-H and sodium aluminosilicate hydrate (N-A-S-H) gels influences the density of the microstructure and mechanical properties of the material (Figure 18b). Silica fume agglomerations are visible as small clusters of silicon and oxygen, which contribute to densification and reduce microstructural porosity (Figure 18b,e). This enhanced matrix density explains the improved mechanical performance of AS mixes matching CS mixes, as the presence of reactive silica promotes stronger aluminosilicate linkages.

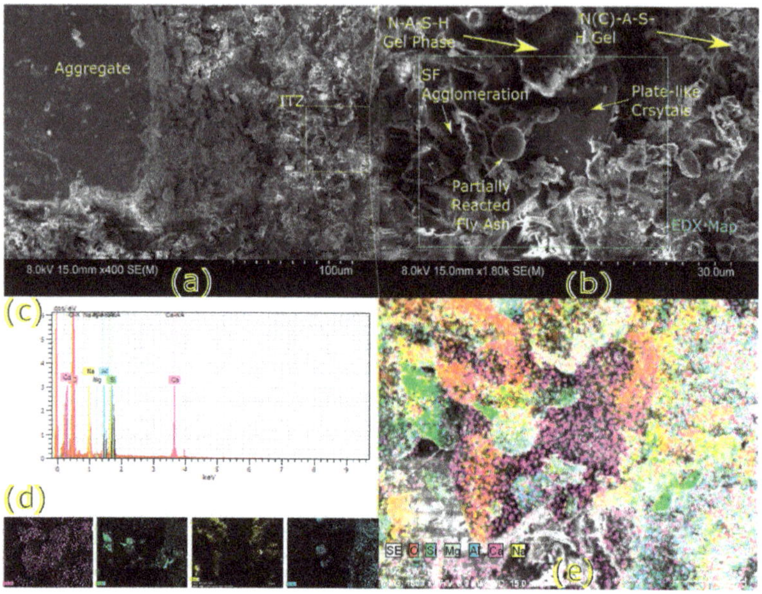

Figure 18. (**a**) SEM Image of AS0.7-80 at magnification of 100 μm, (**b**) 30 μm (**c**) EDX Spectra, (**d**) Elemental maps, (**e**) EDX map.

Figure 19 focuses on specimen AS0.7-80, emphasizing the influence of crystalline phases within the geopolymer matrix. The EDX map and spectra illustrate a similar trend, with calcium-rich regions corresponding to plate-like and needle-like crystalline entities (Figure 19b,e). The surrounding amorphous matrix, consisting of N-A-S-H gels, benefits from the inclusion of silica fume, which ensures a more homogeneous distribution of silicon and aluminum. The reactive silica enhances the geopolymerization process, leading to a dense microstructure with fewer unreacted fly ash particles. The effectiveness of silica fume in balancing crystallinity with matrix densification to achieve superior mechanical properties.

The EDX analysis presented in Table 7 provides crucial insights into the elemental composition and formation of various geopolymer phases. The high Si/Al ratio of 27.1 observed in AS0.5 Point 2 (Figure 20) indicates a dominant N-A-S-H (sodium aluminosilicate hydrate) phase, suggesting that a significant portion of the fly ash underwent effective geopolymerization. This advanced geopolymerization process results in a more refined and interconnected aluminosilicate gel matrix, leading to enhanced mechanical properties. The high Si content in this phase contributes to the overall densification and strength of the geopolymer, as silicon is a critical element in forming strong Si-O-Si and Si-O-Al linkages within the gel network. In contrast, the Ca/Al ratio of 1.38 observed in CS0.5 Point 2 signifies the formation of a C-A-S-H (calcium aluminosilicate hydrate) gel. The presence of a C-A-S-H phase suggests that calcium plays a more dominant role in the geopolymerization process for this mix. However, this often leads to a less homogeneous and more porous microstructure compared to N-A-S-H-dominated systems, which can contribute to reduced mechanical performance. The higher Ca content in C-A-S-H gels generally leads to larger pores and microcracks, resulting in lower compressive and tensile strength compared to N-A-S-H gels.

https://doi.org/10.3390/jcs10010022

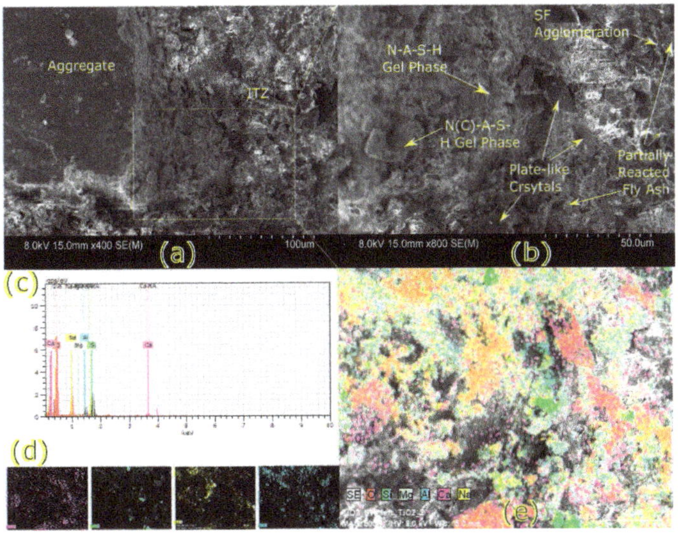

Figure 19. (**a**) SEM Image of AS0.7-80 at magnification of 100 μm, (**b**) 50 μm (**c**) EDX Spectra, (**d**) Elemental maps, (**e**) EDX map.

Table 7. EDS of Specimens AS0.5 and CS0.5.

	Point	O/%	Si/%	Al/%	Fe/%	Na/%	Ca/%	Si/Al	Ca/Al	Ca/Si	Ca/Na
AS0.5	Point 1	48.89	21.55	7.84	9.75	3.44	5.25	2.75	0.67	0.24	1.53
	Point 2	53.9	31.98	1.18	5.44	3.07	1.04	27.1	0.88	0.03	0.34
	Point 3	48.05	18.72	4.91	7.22	7.25	8.15	3.81	1.66	0.44	1.12
CS0.5	Point 1	36.46	5.79	1.38	1.42	9.58	4.51	4.2	3.27	0.78	0.47
	Point 2	23.29	24.94	7.65	7.71	1.99	10.57	3.26	1.38	0.42	5.31
	Point 3	38.79	28.71	6.49	2.13	3.04	1.59	4.42	0.24	0.06	0.52

Figure 20. SEM Image of Specimen AS0.5.

https://doi.org/10.3390/jcs10010022

Moreover, the EDX data shows the presence of partially unreacted fly ash particles in the CS mixes, which corresponds to the lower Si/Al ratios and higher Ca/Na ratios in some points (e.g., CS0.5 Point 3 in Figure 21). This indicates incomplete geopolymerization, where the alkaline activators did not fully react with the fly ash particles. As a result, these unreacted fly ash particles create weak points in the matrix, leading to the formation of microcracks and voids, as observed in the SEM images. This microstructural weakness directly correlates with the reduced mechanical properties seen in the CS mixes.

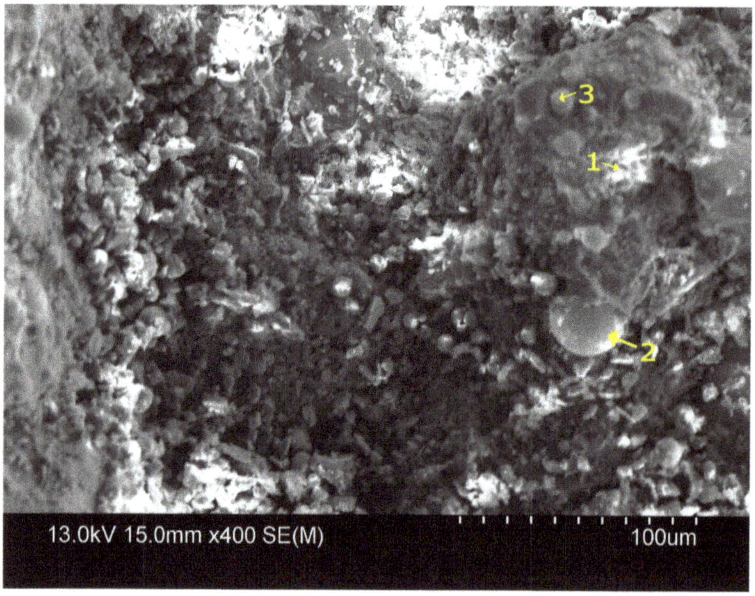

Figure 21. SEM Image of Specimen CS0.5.

This indicates incomplete geopolymerization, where the alkaline activators did not fully react with the fly ash particles. As a result, these unreacted fly ash particles create weak points in the matrix, leading to the formation of microcracks and voids, as observed in the SEM images. This microstructural weakness directly correlates with the reduced mechanical properties seen in the CS mixes.

Figures 22 and 23 present the EDX spectra of the selected target points for AS0.5 and CS0.5, respectively, as indicated in the SEM images in Figures 20 and 21. Table 8 summarizes the corresponding quantitative EDX elemental compositions and calculated elemental ratios for these points.

In contrast, the AS mixes (activated with silica fume) show a more homogeneous and denser microstructure, as evidenced by the consistent formation of N-A-S-H or N(C)-A-S-H phases with higher Si content and fewer unreacted particles. This is because silica fume provides an additional source of reactive silica, promoting the formation of more stable and cohesive geopolymer gels. This contributes to the enhanced compressive strength and reduced porosity observed in AS mixes. Additionally, the significant increase in silicon concentration within the micropores of the fly ash geopolymer confirms the formation of silica (SiO_2) and calcium aluminum silicate (Ca_2Al_2SiO) within the matrix. These phases play a crucial role in reinforcing the geopolymer structure, as they fill the voids and contribute to the overall densification of the matrix. This densification not only enhances the mechanical strength but also improves the durability of the geopolymer, making it more resistant to environmental degradation.

https://doi.org/10.3390/jcs10010022

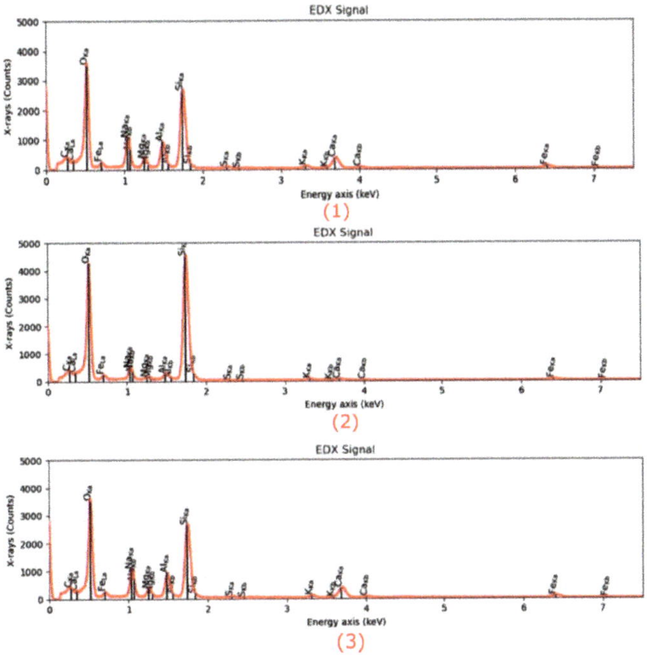

Figure 22. EDX Signal of Points 1, 2, 3 of AS0.5.

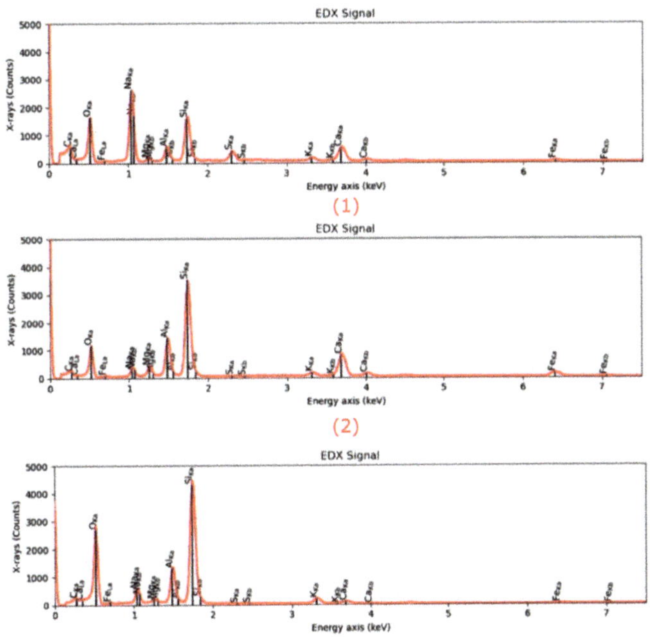

Figure 23. EDX Signal of Points 1, 2, 3 of CS0.5.

https://doi.org/10.3390/jcs10010022

Table 8. Characteristic infrared bands of the samples.

Wavenumber (cm^{-1})	Bonding
1400–1420	Si-O-Si stretching
1050	Asymmetric stretching (O-Si-O)
960–930	Si-O stretching (Si-O-Na)
880–850	Si-O stretching
810–790	Al-O bending vibrations
700–680	Asymmetric stretching (Si-O-Si and Ai-O-Si)

4.4.3. ATR-FTIR Results

ATR-FTIR spectra of the silica fume, fly ash, and geopolymer samples were conducted to study the effect of alkaline activators and the reactions of silicate components (Figure 24). All the significant peaks corresponding to stretching and bending are presented in Table 8. All the geopolymer samples (AS0.5, AS0.7-65, AS0.7-80, CS0.7, CS0.5, AK0.5, and CK0.5) exhibit similar characteristic bands, indicating the occurrence of the expected chemical transformations during geopolymerization. The AS-based and CS-based mixes show comparable stretching and bending shifts, while the AK and CK mixes exhibit slightly smaller shifts, likely due to differences in polymer network structure influenced by potassium.

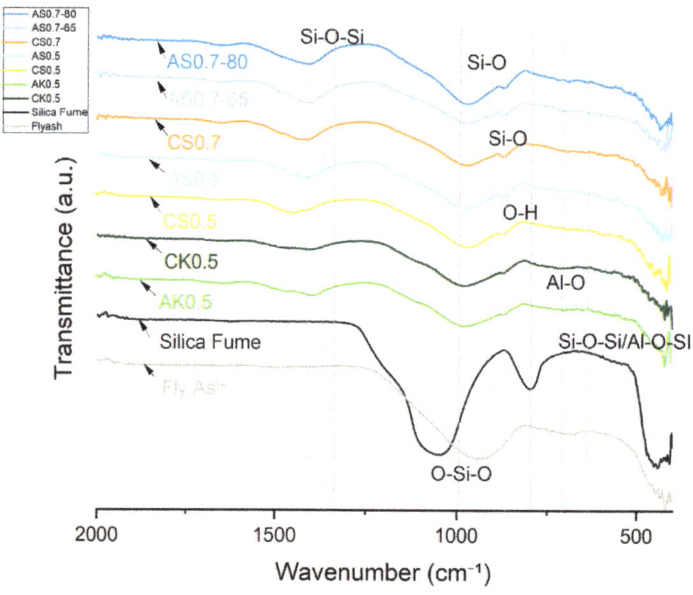

Figure 24. ATR-FTIR spectra of silica fume, fly ash, and geopolymer samples.

ATR-FTIR spectra of silica fume show a broad band at 1050 cm^{-1} and 790 cm^{-1}, indicating O-Si-O asymmetric stretching vibrations [3,25,66]. The fly ash spectra has peaks at 930 cm^{-1} which indicates the Si-O stretching and 690 cm^{-1} which corresponds to asymmetric stretching vibrations of Si-O-Si and Ai-O-Si. The 790 cm^{-1} Al-O peak in the Silica Fume spectra disappears after the application of the alkali activator, and 1050 cm^{-1} shifts to 960 cm-1 for all the geopolymer samples, which may be the result of the O-Si-O bonds changing to Si-O-Al formation [25]. The stretching and bending at 3500 cm^{-1} and 1600 cm^{-1} for water is not present in any of the FTIR spectra, indicating complete removal of hydrates and moisture during the curing process [25]. The 1050 cm^{-1} of the silica fume

https://doi.org/10.3390/jcs10010022

disappears in the geopolymer mix, and there is a ~1400 cm^{-1} peak corresponding to the Si-O-Si stretching in all the geopolymer mixes. The 850 cm^{-1}–880 cm^{-1} indicates the presence of non-bridging oxygen (excess oxygen atoms break the local symmetry of the silica network and create coordinated defects), which results in the silicate species forming bonds with the alkali ions (Na^+) from the alkali activator [70]. These results validate the complete dissolution of silica fume and fly ash with the alkali activator solutions, causing changes in the Si-O bonding and stretching in the geopolymer samples. Table 8 lists the characteristic infrared bands and corresponding bond vibrations of the samples. In addition to peak position shifts, relative changes in band intensity were observed among different mixes, indicating variations in the degree of aluminosilicate network formation. However, full quantitative peak deconvolution, including peak intensity normalization and full width at half maximum (FWHM) analysis, was not performed and is therefore acknowledged as beyond the scope of the present study.

5. Conclusions

This research demonstrates the effectiveness of silica fume-derived sodium silicate alternative (SSA) as an activator for fly ash-based geopolymer mortar through mechanical testing and microstructural analysis using SEM, EDX, and ATR-FTIR. Comprehensive mechanical testing, microstructural characterization, and cost analysis confirmed SSA as a viable, cost-effective, and sustainable alternative for geopolymer applications.

SSA-activated mortars exhibited similar workability to SSC-based mixes while offering extended setting times, providing a practical advantage in construction applications. The mechanical testing revealed that SSA-based geopolymers outperformed SSC-based counterparts in compressive, flexural, and tensile strengths. Notably, the AS0.7-65 mix achieved the highest compressive strength, outperforming its SSC counterpart cured at 80 °C, highlighting the potential to reduce energy consumption without sacrificing performance. Moreover, SSA-based mixes exhibited superior flexural and tensile properties; particularly AS0.7-80, which outperformed all other mixes in bending and tensile performance.

Microstructural analysis via SEM and EDX confirmed that SSA-based geopolymers formed a denser and more homogeneous geopolymer matrix with fewer unreacted fly ash particles. The incorporation of silica fume enhanced the development of the sodium aluminosilicate hydrate (N-A-S-H) gel network, resulting in a well-structured and cohesive matrix. The higher OH^- content in SSA-based mixes facilitated the formation of calcium-rich crystalline phases, further improving strength and ductility. The co-existence of semi-crystalline and amorphous N(C)-A-S-H phases contributed to a more compact microstructure with lower porosity. ATR-FTIR spectra validated these chemical transformations, confirming effective geopolymerization and the development of a robust binder phase.

From an economic perspective, SSA-based geopolymer mortars were approximately 30% more cost-effective than SSC-based ones, primarily due to the lower production cost of SSA compared to commercially manufactured sodium silicate. The sustainability analysis revealed that SSA-based geopolymers reduced CO_2 emissions by approximately 2% compared to SSC-based counterparts. Although this reduction is modest, the improved mechanical performance and cost-effectiveness of SSA significantly enhance its viability for sustainable construction applications. Furthermore, the findings indicate that curing at 65 °C is sufficient for optimal strength development, reducing the need for high-temperature curing and further minimizing the environmental impact.

Despite the advantages of SSA, further research is needed to explore ambient curing methods to enhance sustainability and reduce the energy footprint associated with geopolymer production. Additionally, long-term durability studies should be conducted

https://doi.org/10.3390/jcs10010022

to evaluate the resistance of SSA-based geopolymers to environmental factors such as sulfate attack, freeze–thaw cycles, and carbonation. Investigating the potential of SSA in large-scale applications, including 3D printing and precast elements, would further validate its implementation in the construction industry.

The scalability of SSA production is also an important consideration for industrial adoption. SSA can be synthesized in batch reactors or continuous dissolver systems, both capable of reliably dissolving silica fume in alkaline solutions. Effective temperature and mixing control are required to maintain a stable SiO_2/M_2O ratio and ensure consistent reactivity. The process can be integrated into existing chemical production lines, with manageable energy requirements and standard quality-control procedures. These factors indicate that SSA production can be feasibly scaled for commercial use.

In conclusion, SSA-activated geopolymers present a compelling alternative to traditional SSC-based systems, offering enhanced mechanical properties, improved microstructural integrity, and significant cost benefits. While the environmental impact reduction is modest, the overall performance gains position SSA as a viable and sustainable solution for next-generation geopolymer technology. The results of this study underscore the potential of SSA in advancing geopolymer applications, providing a practical pathway toward more sustainable and durable construction materials.

Author Contributions: N.P.: Writing—review and editing, Writing—original draft, Visualization, Validation, Software, Resources, Project administration, Methodology, Investigation, Formal analysis, Data curation, Conceptualization. P.R.: Formal analysis, Data curation, Writing—original draft. R.G.: Writing—review and editing, Supervision, Project administration, Funding acquisition. All authors have read and agreed to the published version of the manuscript.

Funding: This research received no external funding.

Data Availability Statement: Data will be made available upon request. Available datasets: raw strength data, SEM/EDX data, FTIR spectra, and mix design spreadsheets.

Acknowledgments: This research was conducted at the Camtec Facility at the University of Victoria (UVic). The authors extend their sincere gratitude to Elaine Humphrey and Cristina Cordoba for their invaluable support and expertise at the Advanced Microscopy Facility (AMF), Camtec, UVic. The authors have reviewed and edited the output and take full responsibility for the content of this publication.

Conflicts of Interest: The authors declare no conflicts of interest.

Abbreviations

The following abbreviations are used in this manuscript:

AA	Alkali Activator
AAC	Alkali-Activated Concrete
Al	Aluminum
Al-Si Gel	Aluminum-Silicate Gel
ASTM	American Society for Testing and Materials
AS	Alternative Sodium Silicate mortar
ATR	Attenuated Total Reflectance
A.U.	Arbitrary Unit
C-A-S-H	Calcium Aluminosilicate Hydrate
CAD	Canadian Dollars
CS	Commercial Sodium Silicate mortar
CSH	Calcium Silicate Hydrate
EDX	Energy-Dispersive X-ray Spectroscopy
FA	Fly Ash

https://doi.org/10.3390/jcs10010022

FTIR	Fourier Transform Infrared Spectroscopy
ITZ	Interfacial Transition Zone
KOH	Potassium Hydroxide
MK	Metakaolin
NaOH	Sodium Hydroxide
N-A-S-H	Sodium Aluminosilicate Hydrate
OPC	Ordinary Portland Cement
PC	Portland Cement
PSC	Potassium Silicate (Commercial)
PSA	Potassium Silicate Alternative
RHA	Rice Husk Ash
SEM	Scanning Electron Microscopy
SF	Silica Fume
SSA	Silica Fume-Derived Sodium Silicate
SS	Sodium Silicate
SSC	Sodium Silicate (Commercial)

References

1. Caldas, P.H.C.H.; de Azevedo, A.R.G.; Marvila, M.T. Silica fume activated by NaOH and KOH in cement mortars: Rheological and mechanical study. *Constr. Build. Mater.* **2023**, *400*, 132623. [CrossRef]
2. Santos, T.A.; Cilla, M.S.; Ribeiro, E.D.V. Use of asbestos cement tile waste (ACW) as mineralizer in the production of Portland cement with low CO2 emission and lower energy consumption. *J. Clean. Prod.* **2022**, *335*, 130061. [CrossRef]
3. Emdadi, Z.; Asim, N.; Amin, M.; Ambar Yarmo, M.; Maleki, A.; Azizi, M.; Sopian, K. Development of Green Geopolymer Using Agricultural and Industrial Waste Materials with High Water Absorbency. *Appl. Sci.* **2017**, *7*, 514. [CrossRef]
4. Lemougna, P.N.; Wang, K.; Tang, Q.; Nzeukou, A.N.; Billong, N.; Melo, U.C.; Cui, X. Review on the use of volcanic ashes for engineering applications. *Resour. Conserv. Recycl.* **2018**, *137*, 177–190. [CrossRef]
5. Xia, M.; Nematollahi, B.; Sanjayan, J. Printability, accuracy and strength of geopolymer made using powder-based 3D printing for construction applications. *Autom. Constr.* **2019**, *101*, 179–189. [CrossRef]
6. Billong, N.; Kinuthia, J.; Oti, J.; Melo, U.C. Performance of sodium silicate free geopolymers from metakaolin (MK) and Rice Husk Ash (RHA): Effect on tensile strength and microstructure. *Constr. Build. Mater.* **2018**, *189*, 307–313. [CrossRef]
7. Castillo, H.; Collado, H.; Droguett, T.; Vesely, M.; Garrido, P.; Palma, S. State of the art of geopolymers: A review. *e-Polymers* **2022**, *22*, 108–124. [CrossRef]
8. Ye, H.; Radlińska, A. Fly ash-slag interaction during alkaline activation: Influence of activators on phase assemblage and microstructure formation. *Constr. Build. Mater.* **2016**, *122*, 594–606. [CrossRef]
9. Jwaida, Z.; Dulaimi, A.; Mashaan, N.; Othuman Mydin, M.A. Geopolymers: The Green Alternative to Traditional Materials for Engineering Applications. *Infrastructures* **2023**, *8*, 98. [CrossRef]
10. Furtos, G.; Prodan, D.; Sarosi, C.; Popa, D.; Moldovan, M.; Korniejenko, K. The Precursors Used for Developing Geopolymer Composites for Circular Economy—A Review. *Materials* **2024**, *17*, 1696. [CrossRef]
11. Lopes, A.V.; Lopes, S.M.R.; Pinto, I. Influence of the Composition of the Activator on Mechanical Characteristics of a Geopolymer. *Appl. Sci.* **2020**, *10*, 3349. [CrossRef]
12. Rodríguez, E.D.; Bernal, S.A.; Provis, J.L.; Paya, J.; Monzo, J.M.; Borrachero, M.V. Effect of nanosilica-based activators on the performance of an alkali-activated fly ash binder. *Cem. Concr. Compos.* **2013**, *35*, 1–11. [CrossRef]
13. Bernal, S.A.; Rodríguez, E.D.; Mejia De Gutiérrez, R.; Provis, J.L.; Delvasto, S. Activation of Metakaolin/Slag Blends Using Alkaline Solutions Based on Chemically Modified Silica Fume and Rice Husk Ash. *Waste Biomass Valor.* **2012**, *3*, 99–108. [CrossRef]
14. Tong, K.T.; Vinai, R.; Soutsos, M.N. Use of Vietnamese rice husk ash for the production of sodium silicate as the activator for alkali-activated binders. *J. Clean. Prod.* **2018**, *201*, 272–286. [CrossRef]
15. Raza, M.H.; Khan, M.; Zhong, R.Y. Investigating the impact of alkaline activator on the sustainability potential of geopolymer and alternative hybrid materials. *Mater. Today Sustain.* **2024**, *26*, 100742. [CrossRef]
16. Deabriges, J. Process for the Manufacture of Sodium Silicate. U.S. Patent 4336235A, 22 June 1982. Available online: https://patents.google.com/patent/US4336235A/en (accessed on 18 October 2024).
17. Vinai, R.; Soutsos, M. Production of sodium silicate powder from waste glass cullet for alkali activation of alternative binders. *Cem. Concr. Res.* **2019**, *116*, 45–56. [CrossRef]
18. Sun, Y.; Ghorbani, S.; Dai, X.; Ye, G.; De Schutter, G. Evaluation of rheology and strength development of alkali-activated slag with different silicates sources. *Cem. Concr. Compos.* **2022**, *128*, 104415. [CrossRef]

https://doi.org/10.3390/jcs10010022

19. Cheng, Y.; Cong, P.; Zhao, Q.; Hao, H.; Mei, L.; Zhang, A.; Han, Z.; Hu, M. Study on the effectiveness of silica fume-derived activator as a substitute for water glass in fly ash-based geopolymer. *J. Build. Eng.* **2022**, *51*, 104228. [CrossRef]
20. Billong, N.; Oti, J.; Kinuthia, J. Using silica fume based activator in sustainable geopolymer binder for building application. *Constr. Build. Mater.* **2021**, *275*, 122177. [CrossRef]
21. Oti, J.; Adeleke, B.O.; Mudiyanselage, P.R.; Kinuthia, J. A Comprehensive Performance Evaluation of GGBS-Based Geopolymer Concrete Activated by a Rice Husk Ash-Synthesised Sodium Silicate Solution and Sodium Hydroxide. *Recycling* **2024**, *9*, 23. [CrossRef]
22. Adeleke, B.O.; Kinuthia, J.M.; Oti, J.; Ebailila, M. Physico-Mechanical Evaluation of Geopolymer Concrete Activated by Sodium Hydroxide and Silica Fume-Synthesised Sodium Silicate Solution. *Materials* **2023**, *16*, 2400. [CrossRef] [PubMed]
23. Luna-Galiano, Y.; Leiva, C.; Arroyo, F.; Villegas, R.; Vilches, L.; Fernández-Pereira, C. Development of fly ash-based geopolymers using powder sodium silicate activator. *Mater. Lett.* **2022**, *320*, 132346. [CrossRef]
24. Bellum, R.R.; Muniraj, K.; Madduru, S.R.C. Influence of activator solution on microstructural and mechanical properties of geopolymer concrete. *Materialia* **2020**, *10*, 100659. [CrossRef]
25. Onutai, S.; Osugi, T.; Sone, T. Alumino-Silicate Structural Formation during Alkali-Activation of Metakaolin: In-Situ and Ex-Situ ATR-FTIR Studies. *Materials* **2023**, *16*, 985. [CrossRef] [PubMed]
26. Xi, X.; Zheng, Y.; Zhuo, J.; Zhang, P.; Golewski, G.L.; Du, C. Mechanical properties and hydration mechanism of nano-silica modified alkali-activated thermally activated recycled cement. *J. Build. Eng.* **2024**, *98*, 110998. [CrossRef]
27. Islam, M.T.; Kafle, B.; Al-Ameri, R. Natural Clay in Geopolymer Concrete: A Sustainable Alternative Pozzolanic Material for Future Green Construction—A Comprehensive Review. *Sustainability* **2025**, *17*, 10180. [CrossRef]
28. Con-Fume. Kryton International Inc. 2023. Available online: https://www.kryton.com/technical-info/technical-data-sheets/TDS-Con-Fume.pdf (accessed on 19 September 2024).
29. *ASTM C1240*; Specification for Silica Fume Used in Cementitious Mixtures. ASTM International: West Conshohocken, PA, USA, 2011. [CrossRef]
30. Kang, S.-H.; Hong, S.-G.; Moon, J. Performance Comparison between Densified and Undensified Silica Fume in Ultra-High Performance Fiber-Reinforced Concrete. *Materials* **2020**, *13*, 3901. [CrossRef]
31. Prem, P.R.; Ambily, P.S.; Vankudothu, B.; Bharatkumar, B.H. Sustainable Production of High Performance Concrete. In *Encyclopedia of Renewable and Sustainable Materials*; Elsevier: Amsterdam, The Netherlands, 2020; pp. 527–536. ISBN 978-0-12-813196-1.
32. Lafarge, Centralia Fly Ash. Lafarge Seattle Concrete Lab. 2021. Available online: https://www.seattleconcretelab.net/wp-content/uploads/2022/01/Centralia-Fly-Ash-Cert-1-22-F.pdf (accessed on 19 September 2024).
33. *ASTM C191*; Test Methods for Time of Setting of Hydraulic Cement by Vicat Needle. ASTM International: West Conshohocken, PA, USA, 2021. [CrossRef]
34. Ndlovu, N.Z.N.; Missengue, R.N.M.; Petrik, L.F.; Ojumu, T. Synthesis and Characterization of Faujasite Zeolite and Geopolymer from South African Coal Fly Ash. *J. Environ. Eng.* **2017**, *143*, 04017042. [CrossRef]
35. Patil, K.K.; Allouche, E.N. Impact of Alkali Silica Reaction on Fly Ash-Based Geopolymer Concrete. *J. Mater. Civ. Eng.* **2013**, *25*, 131–139. [CrossRef]
36. Sukmak, P.; Horpibulsuk, S.; Shen, S.-L. Strength development in clay–fly ash geopolymer. *Constr. Build. Mater.* **2013**, *40*, 566–574. [CrossRef]
37. *ASTM C1437*; Test Method for Flow of Hydraulic Cement Mortar. ASTM International: West Conshohocken, PA, USA, 2015. [CrossRef]
38. *ASTM C807*; Test Method for Time of Setting of Hydraulic Cement Mortar by Modified Vicat Needle. ASTM International: West Conshohocken, PA, USA, 2021. [CrossRef]
39. *ASTM C109*; Test Method for Compressive Strength of Hydraulic Cement Mortars (Using 2-in. or [50-mm] Cube Specimens). ASTM International: West Conshohocken, PA, USA, 2020. [CrossRef]
40. *ASTM C307*; Test Method for Tensile Strength of Chemical-Resistant Mortar, Grouts, and Monolithic Surfacings. ASTM International: West Conshohocken, PA, USA, 2023. [CrossRef]
41. *ASTM C293*; Test Method for Flexural Strength of Concrete (Using Simple Beam With Center-Point Loading). ASTM International: West Conshohocken, PA, USA, 2016. [CrossRef]
42. Davidovits, J. *Geopolymer: Chemistry & Applications*, 5th ed.; Institut Géopolymère: Saint-Quentin, France, 2020; ISBN 978-2-9544531-1-8.
43. Rihan, M.A.M.; Onchiri, R.O.; Gathimba, N.; Sabuni, B. Effect of elevated temperature on the mechanical properties of geopolymer concrete: A critical review. *Discov. Civ. Eng.* **2024**, *1*, 24. [CrossRef]
44. De la Peña, F.; Prestat, E.; Lähnemann, J.; Tonaas Fauske, V.; Burdet, P.; Jokubauskas, P.; Furnival, T.; Francis, C.; Nord, M.; Ostasevicius, T.; et al. *Hyperspy, version v2.2.0*; Zenodo: Geneva, Switzerland, 2024. [CrossRef]
45. Ionic radius *Wikipedia*. Available online: https://en.wikipedia.org/w/index.php?title=Ionic_radius&oldid=1218208781 (accessed on 21 February 2025).

46. Sabitha, D.; Dattatreya, J.K.; Sakthivel, N.; Bhuvaneshwari, M.; Sathik, S.A.J. Reactivity, workability and strength of potassium versus sodium-activated high volume fly ash-based geopolymers. *Curr. Sci.* **2012**, *103*, 1320–1327.
47. Sodium Hydroxide. Sigma-Aldrich. Available online: https://www.sigmaaldrich.com/CA/en/product/sigald/221465 (accessed on 19 September 2024).
48. Silica Fume Price Per kg. Henan Superior Abrasives Import & Export Co., Ltd. Available online: https://microsilica-fume.com/silica-fume-price-per-kg.html (accessed on 19 September 2024).
49. Potassium Hydroxide. Sigma-Aldrich. Available online: https://www.sigmaaldrich.com/CA/en/product/sigald/30603 (accessed on 19 September 2024).
50. Potassium Silicate. Thermo Scientific Chemicals. Available online: https://www.thermofisher.com/order/catalog/product/044493.A3.
51. Turner, L.K.; Collins, F.G. Carbon dioxide equivalent (CO_2-e) emissions: A comparison between geopolymer and OPC cement concrete. *Constr. Build. Mater.* **2013**, *43*, 125–130. [CrossRef]
52. Sodium Hydroxide (NaOH). E524. CarbonCloud. Available online: https://apps.carboncloud.com/climatehub/product-reports/id/183314951001 (accessed on 19 September 2024).
53. Potassium Hydroxide (KOH). E525. CarbonCloud. Available online: https://apps.carboncloud.com/climatehub/product-reports/id/1394351136979 (accessed on 19 September 2024).
54. Silica Fume & Sustainability. Norchem, Inc. Available online: https://www.norchem.com/applications-sustainability.html (accessed on 20 September 2024).
55. Oyebisi, S.; Olutoge, F.; Kathirvel, P.; Oyaotuderekumor, I.; Lawanson, D.; Nwani, J.; Ede, A.; Kaze, R. Sustainability assessment of geopolymer concrete synthesized by slag and corncob ash. *Case Stud. Constr. Mater.* **2022**, *17*, e01665. [CrossRef]
56. Fly ash. Concrete Construction Magazine, Fly Ash in Concrete: Availability, Cost, and Market Trends. *Concrete Construction Magazine*. Available online: https://www.concreteconstruction.net (accessed on 20 September 2024).
57. Sand & Gravel Sales. Sharecost Rentals & Sales Ltd. Available online: https://sharecost.ca/bulk.html (accessed on 20 September 2024).
58. Han, Y.; Oh, S.; Wang, X.-Y.; Lin, R.-S. Hydration–Strength–Workability–Durability of Binary, Ternary, and Quaternary Composite Pastes. *Materials* **2021**, *15*, 204. [CrossRef]
59. Zhu, X.; Zhang, Y.; Liu, Z.; Qiao, H.; Ye, F.; Lei, Z. Research on carbon emission reduction of manufactured sand concrete based on compressive strength. *Constr. Build. Mater.* **2023**, *403*, 133101. [CrossRef]
60. Liu, M.; Dai, W.; Jin, W.; Li, M.; Yang, X.; Han, Y.; Huang, M. Mix proportion design and carbon emission assessment of high strength geopolymer concrete based on ternary solid waste. *Sci. Rep.* **2024**, *14*, 24989. [CrossRef]
61. Tushar, Q.; Bhuiyan, M.A.; Abunada, Z.; Lemckert, C.; Giustozzi, F. Carbon Footprint and Uncertainties of Geopolymer Concrete Production: A Comprehensive Life Cycle Assessment (LCA). *C* **2025**, *11*, 55. [CrossRef]
62. Wang, H.; Zhao, X.; Wang, T.; Su, L.; Zhou, B.; Lin, Y. Determination of Gel Products in Alkali-Activated Fly Ash-Based Composites Incorporating Inorganic Calcium Additives. *Adv. Mater. Sci. Eng.* **2022**, *2022*, 7476671. [CrossRef]
63. Barbhuiya, S.; Pang, E. Strength and Microstructure of Geopolymer Based on Fly Ash and Metakaolin. *Materials* **2022**, *15*, 3732. [CrossRef]
64. Lu, Y.; Cui, N.; Xian, Y.; Liu, J.; Xing, C.; Xie, N.; Wang, D. Microstructure Evolution Mechanism of Geopolymers with Exposure to High-Temperature Environment. *Crystals* **2021**, *11*, 1062. [CrossRef]
65. Sivasakthi, M.; Jeyalakshmi, R.; Rajamane, N.P.; Revathi, T. Use of Analytical Techniques for the Identification of the Geopolymer Reactions. *Orient. J. Chem.* **2017**, *33*, 2103–2110. [CrossRef]
66. Fernández-Jiménez, A.; Palomo, A. Composition and microstructure of alkali activated fly ash binder: Effect of the activator. *Cem. Concr. Res.* **2005**, *35*, 1984–1992. [CrossRef]
67. Choi, Y.C.; Park, B. Effects of high-temperature exposure on fractal dimension of fly-ash-based geopolymer composites. *J. Mater. Res. Technol.* **2020**, *9*, 7655–7668. [CrossRef]
68. De Silva, P.; Sagoe-Crenstil, K. Medium-term phase stability of Na2O–Al2O3–SiO2–H2O geopolymer systems. *Cem. Concr. Res.* **2008**, *38*, 870–876. [CrossRef]
69. Cui, H.; Xu, Z.; Guo, T.; Hu, S.; Xia, R.; Zhang, S.; Wang, Y.; Zhou, J. Calcium Hydroxide–Phosphate-Modified Fly Ash Enhances the Adsorption and Stabilization of Soil Lead and Cadmium. *Agronomy* **2024**, *14*, 2905. [CrossRef]
70. Rees, C.A.; Provis, J.L.; Lukey, G.C.; Van Deventer, J.S.J. Attenuated Total Reflectance Fourier Transform Infrared Analysis of Fly Ash Geopolymer Gel Aging. *Langmuir* **2007**, *23*, 8170–8179. [CrossRef]

Disclaimer/Publisher's Note: The statements, opinions and data contained in all publications are solely those of the individual author(s) and contributor(s) and not of MDPI and/or the editor(s). MDPI and/or the editor(s) disclaim responsibility for any injury to people or property resulting from any ideas, methods, instructions or products referred to in the content.

https://doi.org/10.3390/jcs10010022

Journal of
Composites Science

MDPI

Article

Sustainable Mortars Incorporating Industrial Rolling Mill Residues: Microstructural, Physical, and Chemical Characteristics

Ana Laura M. Amorim [1,2], João Victor B. L. Oliveira [2], Rebecca Caroline M. Coelho [1], Bruno S. Teti [1], Esdras C. Costa [1], Nathan B. Lima [3], Kleber G. B. Alves [3] and Nathalia B. D. Lima [1,2,*]

[1] Brazilian Institute for Material Joining and Coating Technologies (INTM), Federal University of Pernambuco, Recife 50740-540, Brazil; analaura.amorim@ufpe.br (A.L.M.A.); rebecca.caroline@ufpe.br (R.C.M.C.); bruno.teti@ufpe.br (B.S.T.); esdras.ccosta@ufpe.br (E.C.C.)

[2] Department of Fundamental Chemistry, Federal University of Pernambuco, Recife 50740-540, Brazil; joao.buregio@ufpe.br

[3] Department of Mechanical Engineering, Federal University of Pernambuco, Recife 50740-530, Brazil; nathan.lima@ufpe.br (N.B.L.); kleber.gbalves@ufpe.br (K.G.B.A.)

* Correspondence: nathalia.blima@ufpe.br

Abstract

New alternatives in the construction industry are essential for economic, sustainable, and environmental progress. In this context, this work investigated three sets of sustainable mortars incorporating industrial lamination waste, assessing their chemical, physical, microstructural, and mechanical properties to inform their development. Cylindrical and prismatic specimens were produced using the following incorporation methods: a reference mortar, mortars with mill scale addition, partial cement replacement with mill scale, and partial sand replacement with mill scale, at proportions of 10%, 20%, 30%, 40%, and 50%. Additionally, analyses including X-ray fluorescence (XRF), X-ray diffraction (XRD), Fourier transform infrared spectroscopy (FTIR), and scanning electron microscopy with energy-dispersive spectroscopy (SEM/EDS) were performed. Physical and mechanical tests, including bulk density, consistency index, capillary water absorption, axial compressive strength, and flexural tensile strength, were also conducted. XRF results indicated an increase in iron oxide content and a decrease in calcium oxide with the addition of mill scale. XRD confirmed the presence of compounds, such as alite and portlandite, which are common in cementitious mortars. FTIR spectra exhibited characteristic functional groups through absorption bands related to Si–O stretching. SEM micrographs revealed slight morphological changes in the composites as the quantity of industrial lamination waste increased, and EDS data supported the XRF findings. The addition of industrial lamination waste affected the spread index and density of the mixtures, while capillary water absorption decreased in some formulations with mill scale. The strength of the mortars increased with the incorporation of industrial lamination waste. In conclusion, using industrial lamination waste in mortars is a technically and environmentally feasible alternative that aligns with the principles of sustainable development and the circular economy in the construction industry.

Keywords: mortars; industrial laminate waste; chemical properties; microstructural properties; mechanical performance

1. Introduction

The transformation of waste into valuable resources for the construction industry aligns with sustainability principles because improving industrial practices is crucial for conserving the planet's natural resources [1,2]. When considering waste generation, it is essential to recognize that rapid and global industrialization has increased production capacity, making goods and services more readily available. This process has helped improve people's quality of life and driven economic growth. However, it also presents a significant challenge: how to manage all this waste effectively [3]. A key example is the steel industry, which plays a vital role in supplying materials to sectors such as shipbuilding, healthcare, transportation, automotive, and civil engineering. Despite its importance to the economy, it is also a significant source of polluting waste. If steel waste is not disposed of or stored correctly, it can cause severe environmental damage [2,4].

Waste generation is not the only sustainability concern. The construction industry accounts for approximately half of all natural resources. As a result, this sector has adopted more sustainable practices, such as the use of industrial waste to produce construction materials, including mortar and concrete. This reduces the need for new raw materials. Utilizing waste to produce construction materials is a practical way to reduce environmental impacts and can also lower costs per square meter. Overall, this approach helps reduce pollution, conserve natural resources, save energy, and decrease waste sent to industrial landfills [5,6].

The metallurgical industry also generates substantial amounts of waste and faces significant challenges in managing it. Even with specific laws in place, many companies ignore these rules, often due to a lack of technical expertise or difficulties in meeting legal requirements [7,8]. Due to its characteristics and properties, steel mill waste has been used to produce recycled construction materials. Managing this waste is one of the most complex and challenging tasks worldwide, as it has a significant environmental impact. For example, scale—a metallurgical waste—can be used in cement production to partially replace clinker, serve as a base for unpaved roads, or act as an aggregate in concrete [2].

While the use of mill scale in concrete production and as a partial replacement for clinker is well documented, recent studies have explored its potential as a fine aggregate in mortars [8–11]. However, most of these studies focus on lower replacement levels or specific properties such as electromagnetic shielding. A comprehensive comparative analysis that evaluates both the chemical effects (as a binder substitute) and the physical packing effects (as an acceptable aggregate substitute) within the same experimental matrix—particularly at high replacement levels (up to 50%)—remains limited. This study bridges this gap by systematically comparing these two incorporation routes, linking mechanical performance directly to microstructural changes (ITZ analysis and densification) to define the optimal role of mill scale in sustainable mortars. However, most of these studies focus on lower replacement levels or specific properties such as electromagnetic shielding [11]. A comprehensive comparative analysis that evaluates both the chemical effects (as a binder substitute) and the physical packing effects (as an acceptable aggregate substitute) within the same experimental matrix—particularly at high replacement levels (up to 50%)—remains limited, with most research focusing on sand replacement [8,9,12]. This study bridges this gap by systematically comparing these two incorporation routes, linking mechanical performance directly to microstructural changes (ITZ analysis and densification) to define the optimal role of mill scale in sustainable mortars. The positioning of this research in relation to the current state of the art is summarized in Table 1.

https://doi.org/10.3390/jcs10010042

Table 1. Benchmarking of recent studies on mill-scale incorporation in cementitious matrices.

Study	Material Replaced	Max. Replacement	Main Research Focus
Pereira et al. [8]	Fine Aggregate	Not specified	Mechanical behavior of concrete using mill scale as aggregate.
Ozturk et al. [11]	Fine Aggregate	15%	Development of mortars for electromagnetic wave shielding.
Parvathikumar et al. [12]	Fine Aggregate	20%	Durability characteristics and river sand replacement.
Siriwattanakarn et al. [9]	Fine Aggregate	100%	Physical, mechanical, and post-fire properties.
Present Study	Cement and Sand	50%	Comparison of chemical vs. physical mechanisms and ITZ analysis.

Furthermore, understanding the upper limits of incorporation is critical for maximizing the environmental benefits of waste valorization. While conservative replacement levels are commonly investigated to ensure safety, they limit the total volume of waste diverted from landfills. By extending the replacement range to 50% and employing advanced microstructural techniques (such as EDS mapping of the Interfacial Transition Zone), this research moves beyond simple feasibility testing. It aims to provide a distinct, mechanism-based explanation for the contrasting behaviors of mill scale as a binder versus an aggregate, thereby offering a scientifically grounded roadmap for the steel and construction industries to implement high-volume recycling strategies effectively [9,10,12].

From a chemical perspective, mill scale is composed of iron oxides, with this composition varying across layers, as depicted in the iron–oxygen (Fe-O) phase diagram. According to Ahmed et al. [6,13], mill scale can be considered a valuable raw material due to its high iron content, low impurity levels, and stable chemical composition. Additionally, the literature indicates that in many cases, mill scale is stored or improperly disposed of in landfills without proper reuse or segregation [6,11,14]. Further research is needed on the potential for reusing mill scale in the construction industry, particularly for mortar production. Thus, incorporating it into cementitious materials presents a promising strategy to reduce environmental impacts, decrease the extraction of natural raw materials, and support practices aligned with the principles of the circular economy.

In this work, sustainable mortars were prepared by replacing cement and sand (at 10%, 20%, 30%, 40%, and 50%) with industrial rolling waste. XRF, FTIR, XRD, SEM, EDS, and both compressive and tensile strength tests were performed.

2. Materials and Methods

2.1. Materials

The raw materials employed to prepare the studied mortars included Portland cement (CPII-Z-32), fine aggregate (fine sand), hydrated lime (CH-1), mill scale from the metallurgical industry (Figure 1), and distilled water. The chemical properties of cement, sand, and lime are detailed in Table 2, while the grain size distribution (granulometry) of both sand and mill scale is presented in Table 3.

https://doi.org/10.3390/jcs10010042

Figure 1. Mill scale: iron oxide residue generated on the steel surface during hot rolling.

Table 2. Chemical compositions of cement, hydrated lime, and sand (unit in % by mass).

	MgO	Al_2O_3	SiO_2	P_2O_5	SO_3	Cl	K_2O	CaO	TiO_2	MnO	Fe_2O_3	SrO	ZrO_3
Cement	1.01	5.48	18.54	0.57	4.25	0.04	1.66	65.18	-	-	3.18	0.09	-
Hydrated lime	0.62	0.37	0.84	0.10	0.14	0.28	0.13	97.77	-	-	-	-	-
Sand	-	1.87	92.17	-	0.14	0.05	0.16	0.63	3.09	-	0.41	-	1.47

Table 3. Particle size distribution of fine aggregate and sand.

Particle Size Distribution					
Sieve Opening (mm)	Test Method	Sand		Mill Scale	
		Percentage Retained (%)	Cumulative Percentage (%)	Percentage Retained (%)	Cumulative Percentage (%)
4.75	NBR 17054	0	0	0.05	0.05
2.36		0.65	0.65	0.10	0.15
1.18		4.21	4.87	4.66	4.81
0.6		24.74	29.60	12.07	16.88
0.3		42.70	72.30	50.58	67.45
0.15		17.66	89.96	17.93	85.38
Background		10.04	100	14.62	100
Total		100	-	100	-

The mill scale used is a byproduct of the steel rolling process and consists predominantly of inorganic iron oxides (Table 4). It was obtained free of organic binders, resins, and hazardous volatile compounds, such as phenols or formaldehyde, ensuring that the study focuses on the interaction between the mineral phases and the cementitious matrix.

Table 4. Chemical composition of mill scale (unit in % by mass).

	Al_2O_3	SiO_2	SO_3	Cl	K_2O	CaO	MnO	Fe_2O_3
Mill scale	0.63	2.26	0.20	0.02	0.09	0.33	0.57	95.90

https://doi.org/10.3390/jcs10010042

The mill scale used in this study was obtained from Gerdau S.A., located in Pernambuco, Brazil. To ensure standardized properties, the residue was oven-dried at 105 ± 5 °C for 24 h to remove moisture, then ground in a ball mill for 30 min at 40 cycles per minute. After processing, a particle size analysis was performed on two 300 g samples in accordance with NBR 17054 [15]. The sieving process utilized mesh openings of 4.75 mm, 2.36 mm, 1.18 mm, 0.6 mm, 0.3 mm, and 0.15 mm, following the same procedure adopted for the fine aggregate. This controlled particle-size distribution enables the scale to function effectively as a microfiller, thereby increasing mass density and improving packing in the composites.

To evaluate the microstructure, morphology, and chemical element mapping of the precursor material, scanning electron microscopy (SEM) and energy-dispersive X-ray spectroscopy (EDS) were performed on the raw mill scale (Figure 2). This characterization is essential for establishing baseline physical and chemical properties of the residue, thereby improving understanding of its interaction with the cementitious matrix. The morphological analysis confirmed the particles' surface roughness and angularity, which are critical factors for mechanical interlocking at the interfacial transition zone (ITZ).

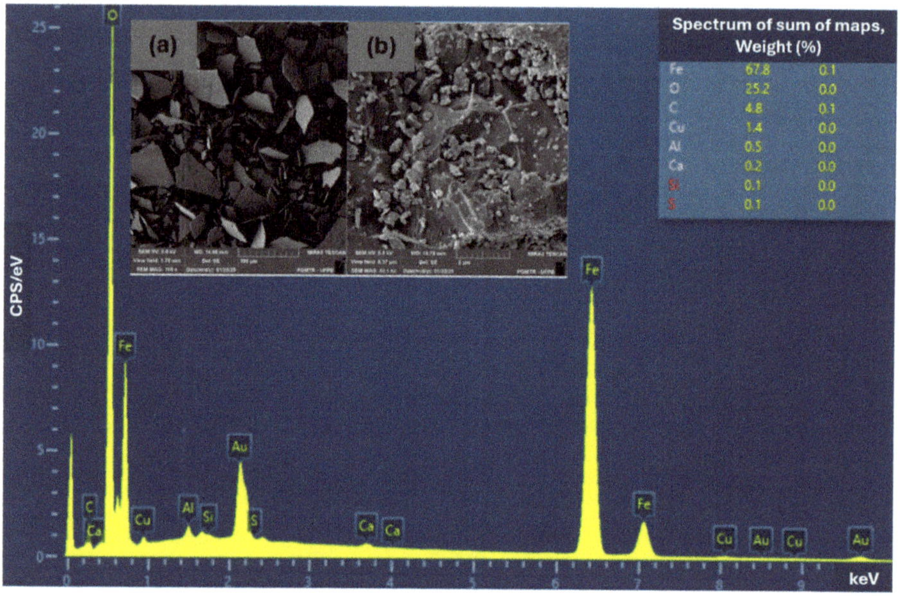

Figure 2. (SEM) of mill scale at different scales and magnifications of (**a**) 500 μm; (**b**) 2 μm, and (EDS).

The morphology and chemical composition of the mill scale are presented in Figure 2. At a 500 μm scale (Figure 2a), the residue consists of predominantly misshapen fragments with markedly rough surfaces. Higher magnification (2 μm) reveals smaller particles adhered to the larger matrix, exhibiting highly irregular surfaces. The EDS spectrum (Figure 2) confirms that Iron (Fe) and Oxygen (O) are the primary constituents, with minor occurrences of Carbon (C), Copper (Cu), Aluminum (Al), Calcium (Ca), and traces of Silicon (Si) and Sulfur (S). This combination of high surface roughness and metallic composition is pivotal for the mechanical interlocking observed in the mortar's interfacial transition zone.

2.2. Proportions of the Mortars

The mortar mix proportions, by volume, were 1:1:6, consisting of Portland cement, CH-I hydrated lime, and sand. Partial replacement was performed under two conditions: (a) partial replacement of cement with scale residue; and (b) partial replacement of sand

https://doi.org/10.3390/jcs10010042

with scale residue in proportions of 10%, 20%, 30%, 40%, and 50%, in both cement and fine aggregate. The water/cement ratio was set at 1.5. While this value may appear high for structural concrete or pure cement mortars, it is characteristic of mixed mortars containing hydrated lime (1:1:6 ratio). Hydrated lime has a high specific surface area, thereby increasing the water demand required to achieve the necessary plasticity for masonry applications. When considering the total binder content (cement + lime), the water-to-binder ratio is approximately 0.70, which falls within the standard range for rendering and masonry mortars, ensuring adequate workability and adhesion. The water-to-cement (w/c) ratio was kept constant at 1.5 across all mixtures to isolate the effect of mill-scale incorporation on mortar properties. This decision ensured that any observed variations in workability, density, and mechanical performance could be directly attributed to the waste's physical and chemical characteristics rather than to fluctuations in water content. This standardization is essential for the reproducibility of the experimental matrix and for a scientifically grounded comparison between the SUB-C and SUB-A series. CP II–Z–32 cement, sand as fine aggregate, CH–I hydrated lime, and scale were used to make the mortars, allowing the impact of adding this residue into the mix to be evaluated. For each formulation, 12 cylindrical and six prismatic specimens will be molded, enabling a detailed analysis of the mortars' mechanical and rheological properties. In this study, mixes with cement substitution are coded as SUB-C, whereas those with substitution in the fine aggregate (sand) are coded as SUB-A. The numerical values that follow each code represent the percentage of cement or sand replaced by scale residue. These abbreviations are consistently used in Table 5.

Table 5. Configuration of the studied mortars.

Code	Mortar Configuration
REF	Reference mortar in a 1:1:6 cement:lime:sand ratio.
SUB-C-10%	Mortar with 10% replacement in a 1:1:6 cement:lime:sand ratio (replacing cement with scale).
SUB-C-20%	Mortar with a 20% replacement in a 1:1:6 cement:lime:sand ratio (replacing cement with scale).
SUB-C-30%	Mortar with a 30% replacement in a 1:1:6 cement:lime:sand ratio (replacing cement with scale).
SUB-C-40%	Mortar with a 40% replacement in a 1:1:6 cement:lime:sand ratio (replacing cement with scale).
SUB-C-50%	Mortar with a 50% replacement in a 1:1:6 cement:lime:sand ratio (replacing cement with scale).
SUB-A-10%	Mortar with a 10% replacement in a 1:1:6 cement:lime:sand ratio (replacing sand with scale).
SUB-A-20%	Mortar with a 20% replacement in a 1:1:6 cement:lime:sand ratio (replacing sand with scale). Mortar with a 30% replacement in a 1:1:6 ratio of cement:lime:sand (replacing sand with scale).
SUB-A-30%	Mortar with a 40% replacement in a 1:1:6 ratio of cement:lime:sand (replacing sand with scale).
SUB-A-40%	Mortar with a 50% replacement in a 1:1:6 ratio of cement:lime:sand (replacing sand with scale).
SUB-A-50%	Mortar with a 20% replacement in a 1:1:6 cement:lime:sand ratio (replacing cement with scale).

2.3. Processing of Mortars

To produce the test specimens (PCs), the following mortar mixes were prepared: a reference mix, a mix with partial replacement of cement, and a blend with partial replacement of sand. Tables 6–8 present the quantities of materials used in the cement-replacement mixes, expressed as percentages of the total. Additionally, a dedicated table will also be provided for the mixes with sand replacement. It is important to note that the water/cement ratio is 1.5 in all proportions.

https://doi.org/10.3390/jcs10010042

Table 6. Materials quantities used in producing reference test specimens.

Reference							
Code	Proportion	Test Specimen	Cement (kg)	Lime (kg)	Sand (kg)	Residue (kg)	Water (kg)
REF	1:1:6	12 cylindrical	0.7312	0.7312	4.387	-	1.023
		6 prismatic	0.6142	0.6142	3.685		0.8599

Table 7. Materials quantities used in preparing cement mass replacement test specimens.

Replacing the Cement Mass							
Code	Proportion	Test Specimen	Cement (kg)	Lime (kg)	Sand (kg)	Residue (kg)	Water (kg)
SUB-C-10%	1:1:6	12 cylindrical	0.6581	0.7312	4.387	0.0731	1.023
SUB-C-20%		6 prismatic	0.5528	0.6142	3.685	0.0614	0.8599
SUB-C-30%	1:1:6	12 cylindrical	0.5850	0.7312	4.387	0.1462	1.023
SUB-C-40%		6 prismatic	0.4914	0.6142	3.685	0.1228	0.8599
SUB-C-50%	1:1:6	12 cylindrical	0.5118	0.7312	4.387	0.2193	1.023
SUB-C-10%		6 prismatic	0.299	0.6142	3.685	0.1842	0.8599
SUB-C-20%	1:1:6	12 cylindrical	0.4387	0.7312	4.387	0.2925	1.023
SUB-C-30%		6 prismatic	0.3685	0.6142	3.685	0.2457	0.8599
SUB-C-40%	1:1:6	12 cylindrical	0.3656	0.7312	4.387	0.3656	1.023
SUB-C-50%		6 prismatic	0.3071	0.6142	3.685	0.3071	0.8599

Table 8. Materials quantities used in preparing sand mass replacement test specimens.

Replacing the Cement Mass							
Code	Proportion	Test Specimen	Cement (kg)	Lime (kg)	Sand (kg)	Residue (kg)	Water (kg)
SUB-A-10%	1:1:6	12 cylindrical	0.7312	0.7312	3.948	0.4387	1.023
SUB-A-20%		6 prismatic	0.6142	0.6142	3.317	0.3685	0.8599
SUB-A-30%	1:1:6	12 cylindrical	0.7312	0.7312	3.510	0.8775	1.023
SUB-A-40%		6 prismatic	0.6142	0.6142	2.948	0.7371	0.8599
SUB-A-50%	1:1:6	12 cylindrical	0.7312	0.7312	3.071	1.316	1.023
SUB-A-10%		6 prismatic	0.6142	0.6142	2.579	1.105	0.8599
SUB-A-20%	1:1:6	12 cylindrical	0.7312	0.7312	2.632	1.755	1.023
SUB-A-30%		6 prismatic	0.6142	0.6142	2.211	1.474	1.023
SUB-A-40%	1:1:6	12 cylindrical	0.7312	0.7312	2.193	2.193	1.023
SUB-A-50%		6 prismatic	0.6142	0.6142	1.842	1.842	0.8599

The specimens were prepared in accordance with the parameters specified in NBR 7215 [16]. The initial step involved separating the cylindrical molds, measuring 5 cm in diameter and 10 cm in height, and weighing out the materials: cement, lime, sand, scale, and water. To facilitate removal from the molds, a liquid release agent was applied to all molds before casting, thereby reducing their release after the material had hydrated. Additionally, the sand was oven-dried to eliminate moisture. Once the materials had been accurately weighed, mortar preparation began. First, cement, lime, and scale (except for the reference mix) were added to the tank, which had been pre-mixed with a spatula. The tank was then attached to the mortar mixer (Figure 3a), and water was added. The mixture was stirred at a low speed

https://doi.org/10.3390/jcs10010042

for 30 s. When the 30 s had elapsed, the mortar mixer beeped, signaling it was time to add the sand. After adding the sand, the mixture was stirred at high speed for an additional 30 s. The mortar was then allowed to rest for 90 s, during which any material adhering to the tank walls and spatula was removed. Subsequently, the equipment was operated at high speed for an additional 60 s, completing the preparation of the cement paste. Once the mortar was ready, specimen molding commenced immediately. Using a spatula, the mortar was distributed in four roughly equal layers, each subjected to 30 uniform blows with a tamper, and the final layer was leveled (Figure 3b). The specimens were then kept in the mold for 24 h; afterward, they were demolded, labeled, and wet-cured until testing.

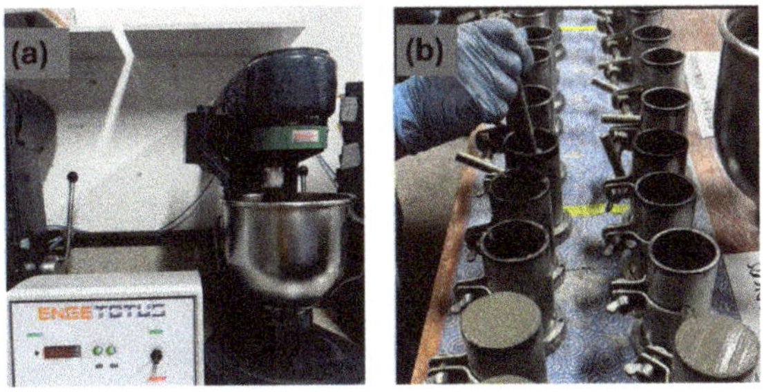

Figure 3. Mortar mixer (**a**); molding of cylindrical test specimens (**b**).

In turn, prismatic specimens were produced in accordance with the parameters specified in NBR 13279 [17]. The first step involved separating the prismatic molds, which measured 4 cm in width, 4 cm in height, and 16 cm in length. A liquid release agent was applied to the molds, and the sand was oven-dried. Additionally, the process performed in the mortar machine was conducted in the same manner.

To begin shaping, the mold was placed on the compaction table, and the initial layer of mortar was evenly added to each compartment of the mold (Figure 4a). Then, 30 drops were applied using the compaction table. Subsequently, the second layer of mortar was applied, repeating the process with an additional 30 drops (Figure 4b). Once these steps were finished, the surface was smoothed with a metal spatula. The specimens remained in molds for 48 h, after which they were demolded, labeled, and wet-cured until the failure date.

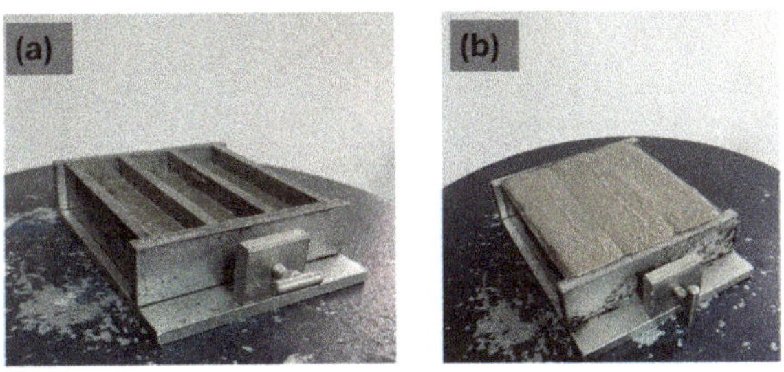

Figure 4. Prismatic molds on the compaction table containing the first mortar layer (**a**); addition and compaction of the second layer (**b**).

2.4. Mechanical Tests

2.4.1. Mass Density Determination

Mass density was determined in accordance with NBR 13278 [18]. To do this, the cylindrical mold, already calibrated and weighed, was filled in three layers of roughly equal height, with 20 blows applied to each layer, for a total of 60 blows per mold. After

https://doi.org/10.3390/jcs10010042

each layer was compacted, the container was dropped three times from an approximate height of 3 cm to further compact the mortar. Excess material was then smoothed with a metal spatula. Finally, the assembly (mold and mortar) was weighed, and its mass was recorded to calculate the mass density using Equation (1).

$$D = \frac{Mc - Mv}{Vr} 1000 \quad (1)$$

where D = mass density (g/cm^3); Mc = mass of the container containing the test mortar (g); Mv = mass of the empty container (g); Vr = volume of the container (cm^3).

2.4.2. Water Absorption by Capillarity

The capillary water absorption test complied with the recommendations of NBR 9779 [19]. Its purpose was to measure capillary rise in the produced mortar samples (CPs) to assess mortar permeability, as increased porosity correlates with a higher void ratio. After a 28-day wet-curing process, the specimens were placed in an oven at 105 ± 5 °C (Figure 5a) and weighed continuously until a stable weight was reached. The CPs were then removed from the oven, air-cooled to 23 ± 2 °C, and weighed to record their initial mass (g). Markings were made on the lower base of each CP at a height of 5 mm. For testing, the CPs were positioned on supports, and the test container was filled with water to maintain a water level 5 ± 1 mm above the bottom face, thereby avoiding contact with other surfaces. During the test, three specimens were used, and their mass was measured at 3, 6, 24, 48, and 72 h of water exposure (Figure 5b). After each measurement, the CPs were promptly returned to the test container.

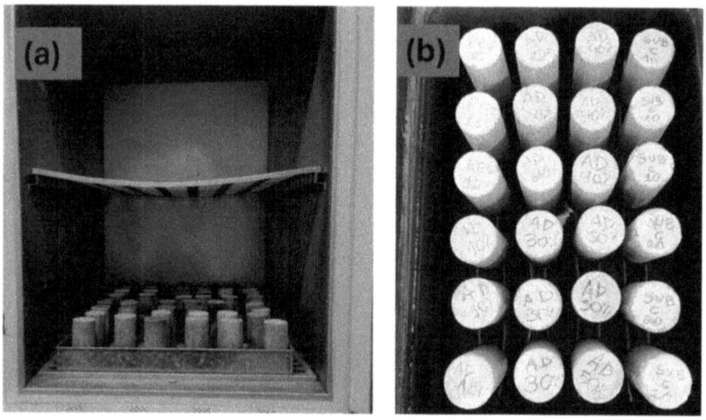

Figure 5. Oven-drying of the mortar test specimens (**a**); performance of the capillary water absorption test on the test specimens (**b**).

After the final weighing, the specimens underwent diametrical compression in accordance with NBR 7222 [20], thereby allowing observation of water rising within them. Absorption was calculated using Equation (2).

$$C = \frac{A - B}{S} \quad (2)$$

where C = is the water absorption by capillarity expressed in (g/cm^2); A = is the saturated mass of the test specimen that remains with one of its faces in contact with the water during the specified period of time, expressed in grams (g); B = is the mass of the dry test specimen, as soon as it reaches a temperature of 23 ± 2 °C, in (g); S = is the cross-sectional area, expressed in square centimeters (cm^2).

https://doi.org/10.3390/jcs10010042

2.4.3. Axial Compression Strength

To verify the material's mechanical behavior, a compression test was conducted. A 20-ton (t) Engetotus manual hydraulic press was used, meeting the requirements established by NBR 7215 [16]. The test was performed on 9 CPs: 3 at 7 days and six at 28 days.

2.4.4. Flexural Tensile Strength

Following the mechanical tests, flexural tensile strength testing commenced, in accordance with the guidelines outlined in NBR 13279 [17]. An Engetotus electric press with a maximum capacity of 100 tons was used for this purpose. Three 7-day-old and three 28-day-old CPs were tested. After flexural rupture, each half was subjected to compression testing to allow for further analysis of the material's strength.

2.4.5. Determining the Consistency Index (Flow Table)

The flow table test evaluates the mortar's spreadability (fluidity) and helps analyze its workability in accordance with NBR 13276 [21]. For each proportion tested, a test was conducted to assess the effect of adding or replacing the scale in the mortar.

After preparing the mortar on a clean table, the truncated cone mold was filled and positioned centrally on the table (Figure 6b) to determine the consistency index. Before this step, the flow table apparatus used in the test is shown in Figure 6a. The mold should be held firmly and filled in three successive layers of about equal height. Each layer was compacted with the socket using the specified number of blows: 15 for the first layer, 10 for the second, and 5 for the third. The blows were evenly distributed to ensure proper compaction. Once the mortar was compacted, the mold was removed, and the mortar before spreading is shown in Figure 6c. The crank was then immediately rotated, delivering 30 strokes in 30 s. Finally, after the spreading was complete, the spread is shown in Figure 6d, and three measurements were taken at different points to calculate the average spread diameter.

Figure 6. Flow table used for consistency testing (**a**); introduction of the mortar into the truncated-cone mold (**b**); mortar surface before the flow test (**c**); mortar spread after lifting the mold (**d**).

https://doi.org/10.3390/jcs10010042

2.5. Microstructural Characterization

2.5.1. X-Ray Fluorescence (XRF)

X-ray fluorescence (XRF) was employed to determine the composition of the precursor materials and the resulting cementitious composites. This analysis was performed using a Rigaku Primini (Rigaku, Kyoto, Japan) instrument operating at 40 kV and 1.25 mA. The samples were prepared as pellets in a press at a force of 10 tons (tf). For scale, boric acid (H_3BO_3) was used as a binding agent to prepare the pellets.

2.5.2. X-Ray Diffraction (XRD)

To analyze the microstructure of the scale and mortar, samples from the residue and each proportion were separated and subjected to X-ray Diffraction (XRD) testing. The samples were dried in an oven at 105 ± 5 °C for 24 h to remove moisture, then ground in a porcelain mortar and pestle. After this process, a portion of the sample was set aside for XRD testing. The samples were analyzed using a Rigaku SmartLab X-ray diffractometer (Rigaku, Kyoto, Japan) at an operating voltage of 40 kV, a current of 30 mA, a 2θ range of 5–80°, a scanning speed of 2°/min, and a sampling pitch of 0.02°.

2.5.3. Fourier Transform Infrared Spectroscopy (FTIR)

Fourier transform infrared spectroscopy (FTIR) was employed to identify the chemical bonds and molecular interactions present in the samples. The analysis was performed with a Bruker Alpha II spectrometer (Bruker, Billerica, MA, USA) using the universal transmission module. Sample preparation involved forming pellets, which were then placed in the sample holder. To do this, the sample was mixed with potassium bromide (KBr) and compressed in a Specac press (Bruker, Massachusetts, USA) under a 10-tonne load (10 tf) to form the pellet. Spectra were recorded in the 4000–600 cm^{-1} range in transmittance mode, with a resolution of 4 cm^{-1}, for a total of 64 measurements.

2.5.4. Scanning Electron Microscopy (SEM/EDS)

Microstructural analysis was conducted using Scanning Electron Microscopy (SEM) (Tescan, Brno, Czech Republic). Samples of each material—such as scale, reference mortar, and other proportions of addition and replacement—were prepared for this purpose. The samples were metallized before SEM analysis. A Tescan MIRA 3 microscope, equipped with an energy-dispersive X-ray analyzer (EDS) (Tescan, Brno, Czech Republic), an FEG filament, and secondary electron detectors, was used for imaging. Images were captured at an accelerating voltage of 5 kV. For chemical analysis, an energy-dispersive X-ray spectrometer (EDS) (Ultim Max, Oxford Instruments, Oxford, UK) was employed in high-vacuum chamber mode at an accelerating voltage of 15 kV.

2.6. Statistical Analysis

The experimental results were subjected to a statistical analysis to verify the significance of the differences observed between the reference mortar and the modified mixtures. A One-Way Analysis of Variance (ANOVA) was performed, followed by Tukey's post hoc test, with a significance level set at 5% ($p < 0.05$). The error bars presented in the graphs represent the standard deviation of the mean values obtained from the tested specimens ($n = 12$ for cylindrical and $n = 6$ for prismatic specimens).

3. Results and Discussions

3.1. Physical Aspects

Initially, the mortar consistency index was evaluated in accordance with NBR 13276 [21] (flow table test) (Figure 7), indicating that partial replacement of cement with

https://doi.org/10.3390/jcs10010042

mill scale maintained or slightly increased the mortar's workability. The highest spread rates were recorded at 30% and 50% replacements, with values of 255 mm and 257 mm, corresponding to increases of 4.51% and 5.33%, respectively, compared to the reference sample. Unlike cement, which is highly hygroscopic and reacts rapidly with water, mill scale is composed of iron oxides that do not react quickly with water. This slow reactivity of iron oxides helps keep the mortar's fluidity [22].

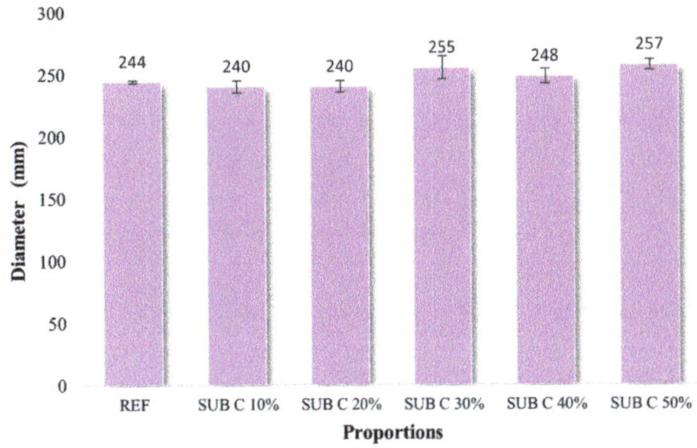

Figure 7. Consistency Index (flow table spread in mm) of mortars incorporating mill scale residue as a partial cement replacement.

As shown in Figure 8, the SUB A 30% formulation exhibited a localized deviation from the overall downward trend, with a higher standard deviation. This behavior is likely linked to the irregular and angular morphology of the mill scale fragments (as shown in SEM, Figure 2), which increases internal friction. At the 30% threshold, the balance between the non-absorbent nature of the scale—which increases the available free water—and the physical interlocking of the metallic particles may lead to greater experimental variability in the flow table test. At higher concentrations (40% and 50%), the physical packing and the high density of the residue (5.5 g/cm^3) clearly override the effect of free water, resulting in a more consistent reduction in workability.

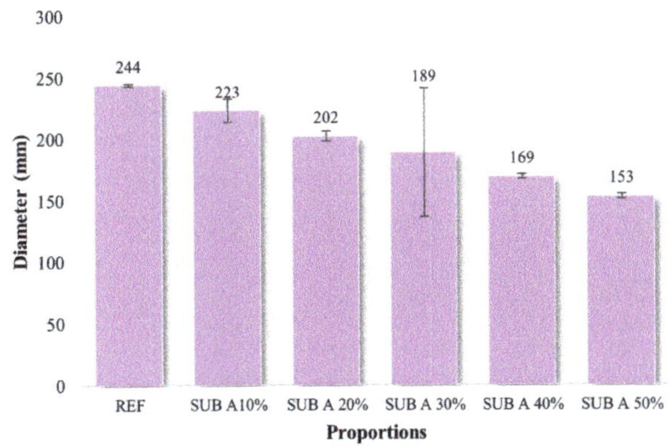

Figure 8. Consistency Index (flow table spread in mm) of mortars incorporating mill scale residue as a partial sand replacement.

From Pereira, Verney, and Lenz [8] and the results obtained, the scale requires an increase in water content to maintain workability, as observed. However, all incorporation and proportion experiments in this study were conducted at a w/c ratio of 1.5.

https://doi.org/10.3390/jcs10010042

Maintaining a constant w/c ratio of 1.5 directly influenced these rheological results. Since mill scale particles are predominantly metallic and non-absorbent, replacing natural sand with this residue theoretically increases the volume of free water available in the paste. However, the drastic reduction in the consistency index observed for the SUB-A group suggests that the physical properties of the residue—such as its high mass density (5.5 g/cm^3), irregular morphology, and surface roughness—exerted a greater influence on internal friction than the presence of additional free water. This mechanism also explains why mechanical strength was maintained in the SUB-A series: effective physical interlocking and dense packing of the residue particles compensated for the high water content of the mortar matrix.

Figure 9 shows the density of the mortars, illustrating a gradual, consistent increase in mass density with increasing scale content. The initial density of the reference mortar was 2250 kg/m^3, and substitutions of 10%, 20%, and 30% increased the density by 0.89%, 1.78%, and 2.22%, respectively. Beyond 40%, density decreased by 1.33% relative to the reference value; at 50%, it stabilized and returned to its initial value.

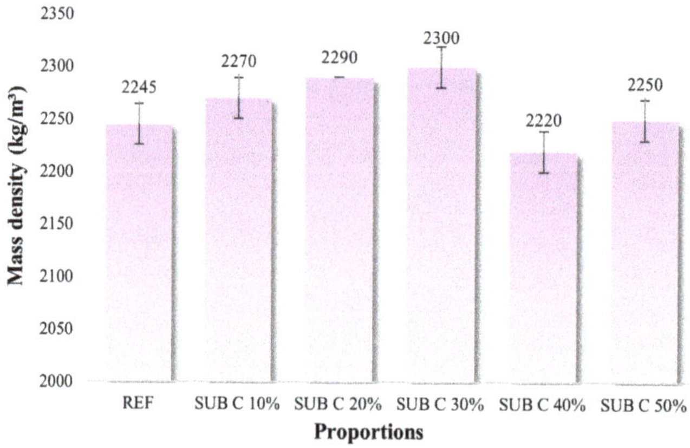

Figure 9. Mass density of mortars incorporating mill scale residue as a partial cement replacement.

As shown in Figure 10, a steady and consistent increase in mass density was observed with increasing scale content. The initial density of the reference mortar was 2250 kg/m^3, establishing a baseline for comparison with the REF (2250 kg/m^3). The increases were approximately 2.22%, 4.44%, 5.33%, 8.44%, and 10.67%, respectively.

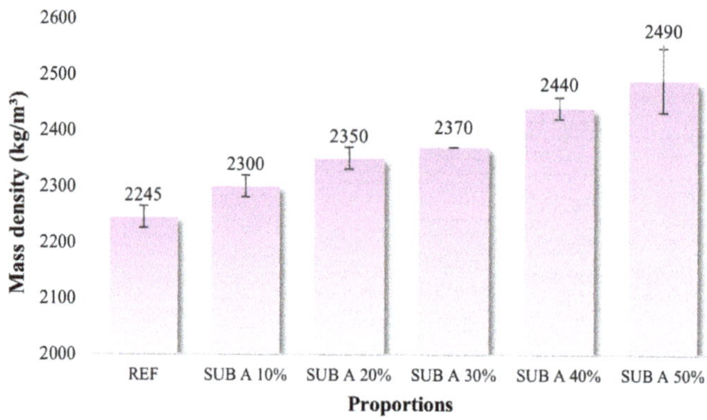

Figure 10. Mass density of mortars incorporating mill scale residue as a partial sand replacement.

https://doi.org/10.3390/jcs10010042

The behavior observed in sustainable mortars shows distinct effects on the compaction and internal structure of the cement matrix. According to Gagliotti [6], the scale has a density of 5.5 g/cm^3, typical of dense metallic and mineral materials. This characteristic largely accounts for the increase in mortar mass density, particularly when sand is partially replaced by scale.

When replacing sand, which has a density of about 2.6–2.7 g/cm^3, adding scale doubles the density. This can improve mechanical properties, such as increased strength and lower permeability, as the matrix becomes less porous. Omrane and Rabeni [23] suggest a direct link between density, porosity, and strength. In the specific case of the mortars studied here, the increase in bulk density (Figures 9–11) is primarily a result of the high specific gravity of the mill scale (5.5 g/cm^3) compared to the sand it replaces. While in conventional materials a higher density may directly indicate lower porosity, in these composites the density increase is a physical consequence of the heavier metallic inclusions. Therefore, the actual refinement of the pore network and the reduction in effective porosity are better evidenced by the capillary water absorption results (discussed in Section 3.2), rather than the bulk density alone.

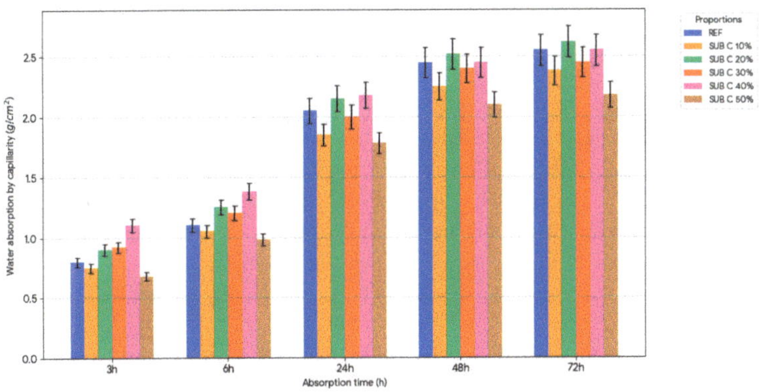

Figure 11. Capillary water absorption as a function of time for mortars incorporating mill scale residue as a partial cement replacement.

The proportions with 10% and 50% replacement showed water absorption values that were lower than or very close to the REF at all time points analyzed. However, the formulations with 20%, 30%, and 40% replacement exhibited absorption values higher than the REF, suggesting increased porosity or greater heterogeneity of the cement matrix, possibly due to an imbalance in the cement-to-aggregate ratio that compromised proper compaction. This increased absorption could adversely affect durability by facilitating penetration by aggressive agents.

Based on Figure 12, all mortars with sand replacement had lower absorption values than REF, which showed the highest absorption throughout the test. Among the different proportions, the formulations with 20% and 30% replacement exhibit the lowest water absorption values.

Figure 12 suggests that replacing sand with scale may have reduced capillary porosity in the mortars, likely due to the filling effect of the fine residue particles, which promotes a denser matrix that is less permeable to moisture. Additionally, the 10%, 40%, and 50% formulations outperformed REF, although their absorption values were slightly higher than those previously reported. This behavior could be linked to excess residual material, which at high levels can promote the formation of unwanted pores, even if it does not exceed the REF values.

https://doi.org/10.3390/jcs10010042

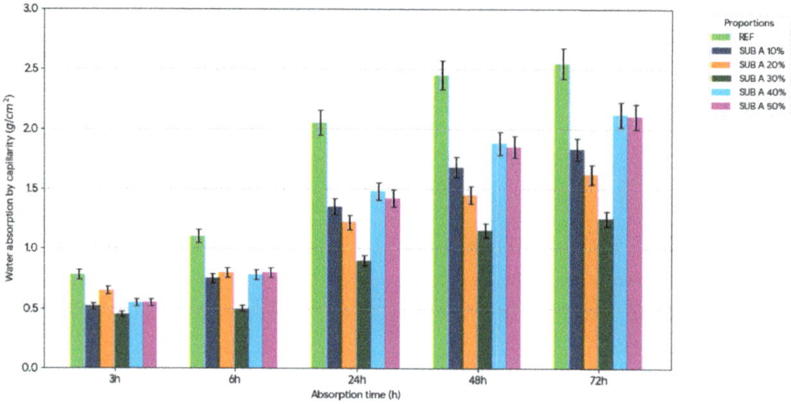

Figure 12. Capillary water absorption as a function of time for mortars incorporating mill scale residue as a partial sand replacement.

In summary, each incorporation led to distinct behaviors; however, some similarities emerged across the various compositions. In all proportions, water absorption increased over time, as expected in cementitious materials. Additionally, across all compositions, some proportions outperformed the reference, exhibiting lower water absorption values. It is essential to note that replacing cement, a crucial component in the formation of C–S–H and portlandite, can adversely affect microstructural development, particularly at intermediate levels. According to Mehta and Monteiro [24] and Taylor [25], the proper formation of hydration products (C–S–H) relies on the presence of reactive cement, and excessive replacements can dilute the matrix and increase permeability.

3.2. Mechanical Behavior

Figure 13 presents the compressive strength results expressed as mean values ± standard deviation. Statistical analysis (ANOVA) followed by Tukey's post hoc test confirmed a significant reduction ($p < 0.05$) in compressive strength for all replacement levels compared to the reference (REF). As indicated by the distinct grouping letters (a–e) in Figure 13, the decrease was progressive and statistically significant with each mill-scale increment, reaching a minimum at 50% replacement. Portland cement primarily forms hydration products, such as C–S–H, portlandite, and ettringite, which together provide mechanical strength [24]. Therefore, it can be concluded that scale lacks the same binding capacity or hydration potential as Portland cement, thereby compromising the final mechanical strength.

Although a reduction in compressive strength was observed, this behavior is consistent with the literature on mortars incorporating other waste materials, such as expanded polystyrene (EPS) or recycled plastics, in which strength losses are accepted in exchange for environmental benefits [10,26]. Despite the reduction, the mortars with intermediate cement replacement levels still exhibit sufficient strength for non-structural applications, such as masonry coating or partition blocks.

For mortars with partial sand replacement (Figure 14), the statistical analysis (ANOVA/Tukey) indicated no significant difference ($p > 0.05$) between the REF sample and the remaining formulations at 28 days, as indicated by the shared grouping letter 'a' in the chart. This statistical analysis supports the observation that replacing sand with mill scale, even at higher levels, preserves the mechanical integrity of the matrix. Although minor fluctuations in mean values were observed (such as the slight decrease at 30% followed by recovery at 50%), they fall within the margin of error (standard deviation), confirming that the mill scale acts effectively as a filler without disrupting matrix cohesion. Unlike cement

https://doi.org/10.3390/jcs10010042

replacement, sand does not affect hydration processes and primarily acts as a filler [24]. Therefore, replacing sand with an inert material, such as scale, does not cause such notable interference. Ozturk et al. [8] also examined the effects of using scale as a partial replacement for fine aggregate; in this study, they found that at levels up to 15%, both compressive and flexural tensile strengths increased. However, levels above 15% reduced mechanical properties due to the formation of irregular structures and agglomerates that impaired the compactness and adhesion of the cement paste.

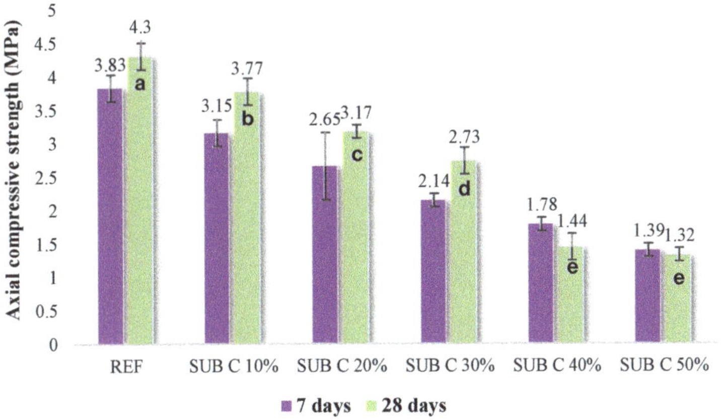

Figure 13. Axial compressive strength of the reference mortar and of mortars incorporating mill scale residue as a partial cement replacement. Note: Means followed by different lowercase letters indicate significant differences at 28 days according to Tukey's post hoc test ($p < 0.05$).

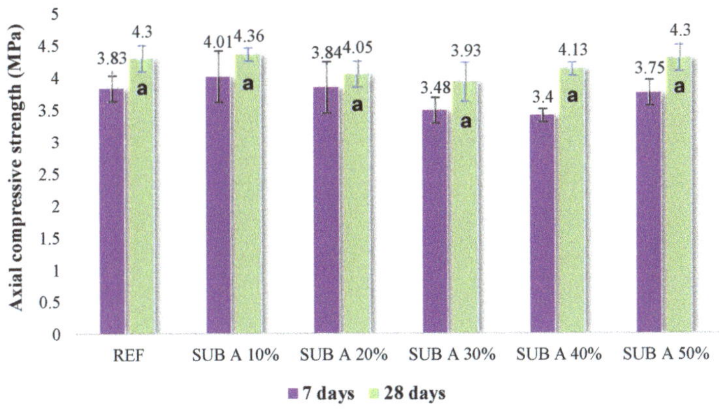

Figure 14. Axial compressive strength of the reference mortar and mortars incorporating mill scale residue as a partial sand replacement. Note: The shared lowercase letter 'a' indicates that there are no statistically significant differences ($p > 0.05$) between the mixtures at 28 days according to Tukey's post hoc test.

Based on the axial compressive strength test, the scale's performance varies with the method of incorporation. When scale is used as a partial cement substitute, mechanical strength gradually decreases, indicating that scale neither promotes agglomeration nor contributes to the hydration products required for matrix cohesion. However, when the residue is used as a partial sand substitute or even added directly, scale is technically feasible (depending on the proportion added), maintaining or even slightly enhancing compressive strength at moderate levels without disrupting chemical hardening processes [27].

https://doi.org/10.3390/jcs10010042

The flexural tensile strength results for cement replacement (Figure 15) align with the axial compression findings. Statistical analysis (ANOVA/Tukey) confirmed that the reduction in tensile strength is significant ($p < 0.05$) for replacement levels above 10%, as shown by the distinct grouping letters (a–d) in Figure 15. This indicates that the loss of hydration products, such as C–S–H, directly impacts the matrix's capacity to resist tensile stresses. Despite decreases in strength values, all proportions showed gains in strength between 7 and 28 days, indicating that hydration and strength development continue, albeit less efficiently due to the absence of the material responsible for the hydration reactions. Additionally, Figure 15 displays the flexural tensile strength of the reference mortar and the mortar with cement replaced by scale.

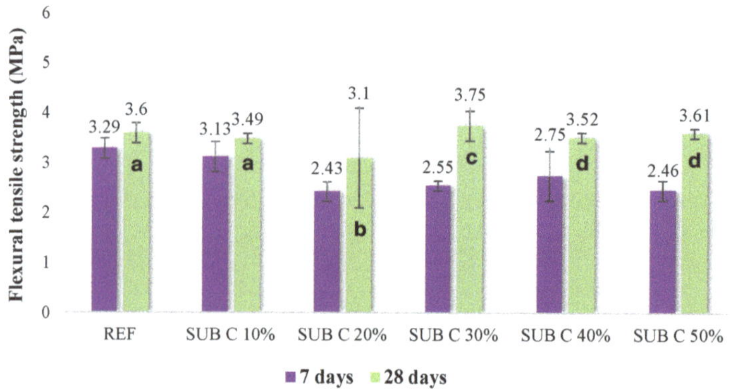

Figure 15. Flexural tensile strength of mortars for SUB-C series. Note: Means followed by the same lowercase letters for the same curing age are not significantly different ($p > 0.05$) according to Tukey's post hoc test.

For the sand-replacement group (Figure 16), the ANOVA test followed by Tukey's post hoc test showed no statistically significant loss of strength relative to the reference ($p > 0.05$) at all tested replacement levels (up to 50%), as indicated by the shared grouping letter 'a'. In fact, formulations like SUB A 20% and SUB A 40% showed mean values numerically higher than the REF. However, these increases fall within the statistical margin of equivalence, confirming that mechanical performance is maintained. Replacing sand with scale did not directly impact the hydration products, allowing the matrix to develop correctly.

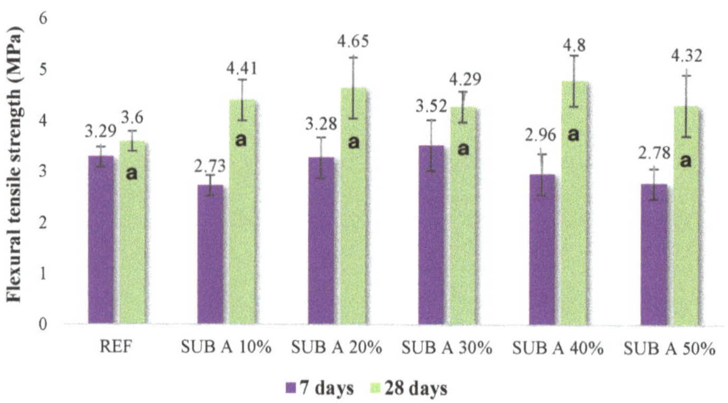

Figure 16. Flexural tensile strength of mortars with partial sand replacement (SUB-A). Note: Means followed by the same lowercase letter (a) for the same curing age do not differ significantly according to Tukey's post hoc test ($p > 0.05$).

https://doi.org/10.3390/jcs10010042

Furthermore, the flexural tensile strength also supports the results of the axial compression test, in which mortars with up to 50% sand replacement either maintained or exceeded the reference values. Therefore, the results show that using scale as a partial substitute for sand in mortars is feasible, effectively preserving mechanical properties throughout the tested range.

Subsequently, the compressive strength test of the prismatic CPs was performed (Figures 17 and 18).

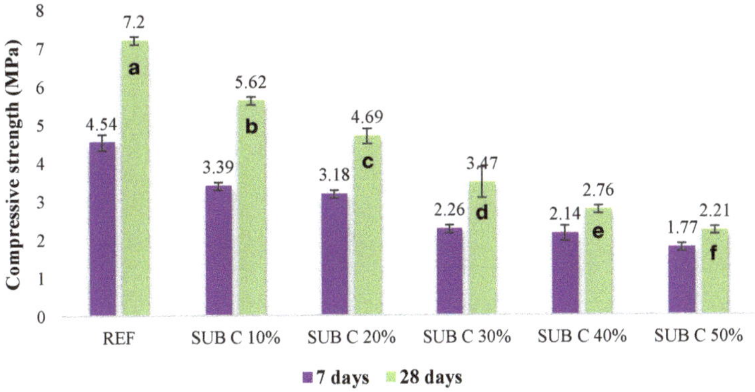

Figure 17. Compressive strength of prismatic specimens for SUB-C series. Note: Means followed by different lowercase letters indicate significant differences at 28 days according to Tukey's post hoc test ($p < 0.05$).

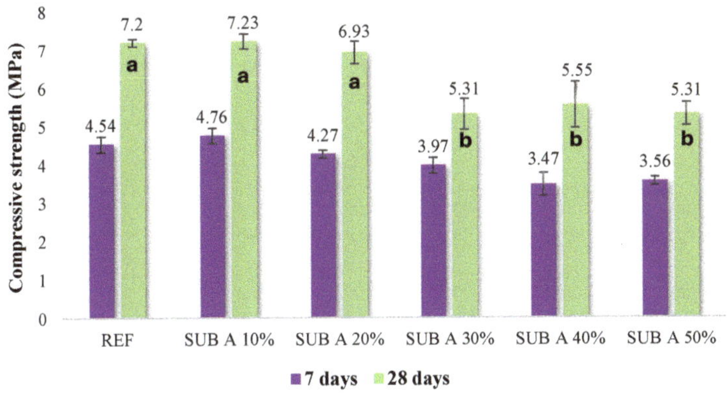

Figure 18. Compressive strength of prismatic specimens for SUB-A series. Note: Means followed by different lowercase letters (a, b) indicate a statistically significant difference ($p < 0.05$) between the replacement levels at 28 days according to Tukey's test.

The results obtained from prismatic specimens (Figures 17 and 18) provide further validation. Statistical testing on the SUB C group (Figure 17) confirms a significant and progressive reduction ($p < 0.05$) in strength corresponding to the cement dilution, as evidenced by the distinct grouping letters (a–f). Conversely, for the SUB A group (Figure 18), the analysis shows that compressive strengths for 10% and 20% replacements are statistically equivalent to the reference (letter 'a', $p > 0.05$). However, for levels from 30% to 50%, a significant decrease was observed (letter 'b'), although these values remained within safe limits for non-structural applications. This supports the proposal's feasibility from a sustainability perspective, without compromising the safety of the materials used in civil construction.

https://doi.org/10.3390/jcs10010042

3.3. Chemical Aspects of the Mortars

In the context of mortars with partial replacement of cement by scale (Table 9), the results showed changes in the principal oxides, such as CaO, Fe_2O_3, and SiO_2. As the replacement percentage increased, the Fe_2O_3 content rose significantly, from 1.60% to 14.51%, reflecting the high iron oxide content in the scale. Conversely, the CaO content steadily decreased from 61.85% in the reference sample to 49.41%. Additionally, there was a slight increase in SiO_2, likely due to the presence of silicate phases in the scale.

Table 9. Chemical composition of mortars with partial cement replacement by scale (wt%).

	CaO	SiO_2	Al_2O_3	Fe_2O_3	MgO	P_2O_5	SO_3	Cl	K_2O	TiO_2	ZrO_2
REF	61.85	29.07	3.50	1.60	0.90	0.40	1.62	0.05	0.43	-	058
SUB C 10%	57.81	29.40	3.14	4.30	0.86	0.35	1.61	0.05	0.34	1.63	0.50
SUB C 20%	55.79	27.59	2.92	7.97	0.81	0.36	1.45	0.09	0.80	1.69	0.52
SUB C 30%	53.21	29.14	2.77	11.19	0.67	0.33	1.29	0.07	0.77	-	0.54
SUB C 40%	50.31	29.24	2.64	12.91	0.60	0.35	1.19	0.08	0.57	1.58	0.54
SUB C 50%	49.41	30.67	2.55	14.51	0.67	0.26	1.09	0.09	0.21	-	0.52

Table 10 shows the results for mortars with partial sand replacement by scale, in which a similar trend to that observed with the addition of the residue can be observed, although much more expressive, with a gradual increase in relation to Fe_2O_3 and a reduction in CaO. Another relevant point is the behavior of SiO_2, which shows a gradual, though more discreet, reduction, from 29.07% to 16.05%.

Table 10. Chemical composition of mortars with partial sand replacement by scale (wt%).

	CaO	SiO_2	Al_2O_3	Fe_2O_3	MgO	P_2O_5	SO_3	Cl	K_2O	MnO	ZrO_2
REF	61.85	29.07	3.50	1.60	0.90	0.40	1.62	0.05	0.43	-	0.58
SUB A 10%	50.39	25.39	3.10	16.97	0.88	0.36	1.51	0.08	0.95	-	0.35
SUB A 20%	44.91	22.94	3.09	25.76	0.85	0.34	1.39	0.04	0.37	-	0.29
SUB A 30%	39.58	20.52	3.08	33.47	0.94	0.32	1.34	0.03	0.30	0.21	0.19
SUB A 40%	35.55	18.71	2.86	39.77	0.82	0.32	1.23	0.03	0.28	0.26	0.15
SUB A 50%	32.61	16.05	2.56	45.72	0.87	0.26	1.15	0.05	0.32	0.27	0.12

3.4. Structural Characteristics of the Sustainable Mortars

Figure 19 displays the diffractograms of the reference mortars and those with partial cement replacement by scale after 28 days of curing. The crystalline phases identified were identical to those previously reported and exhibited approximately the same scattering angles.

The intensity of the Portlandite (P) peaks gradually decreases as cement scale replacement increases, particularly at 40% and 50% replacement levels. This decline indicates reduced formation of hydration products because less cement is available for reaction. Similarly, the alite (C_3S) peak shows a decreasing intensity with increasing scale content, supporting the idea of reduced reactive potential in the cementitious system.

Figure 20 presents the diffractograms of reference mortars and formulations with varying levels of partial replacement of sand by scale after 28 days of curing. Overall, it is evident that replacing sand with scale yields diffractograms that are similar to those obtained with other substitutions, showing the same crystalline phases: portlandite (P), alite (C_3S), calcite (C), and quartz (Q). The peaks remain in consistent positions, with only the intensities of the quartz (Q) peaks changing with the scale proportion used.

https://doi.org/10.3390/jcs10010042

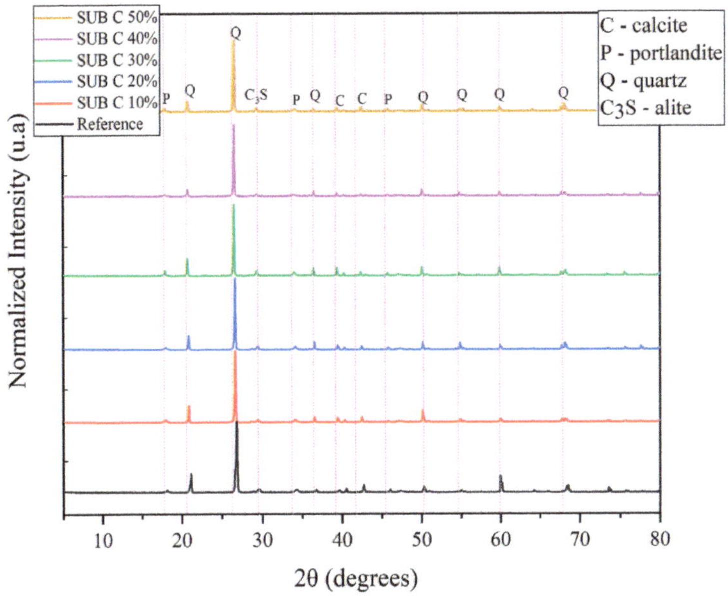

Figure 19. XRD patterns of the reference mortars and those with cement replaced by scale.

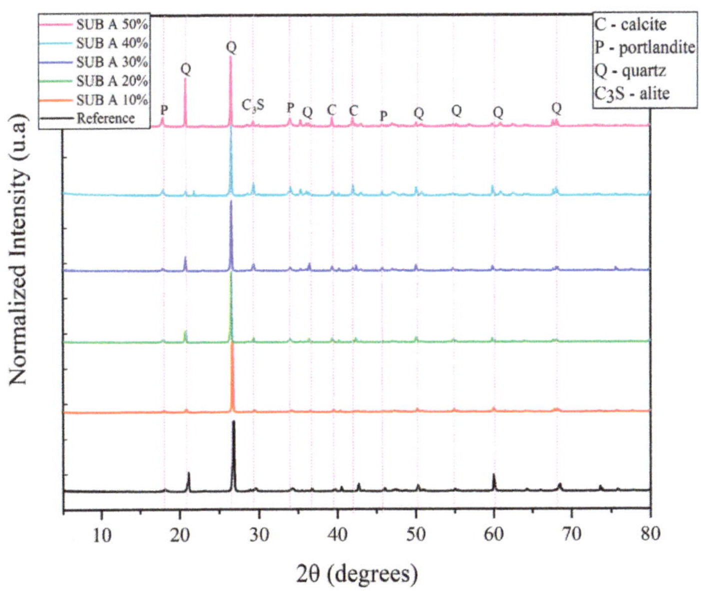

Figure 20. XRD of the reference mortars and those with sand replaced by scale.

Figure 21 presents the spectral results for mortars with partial cement replacement at different scales. The same absorption bands observed in the previous spectrum are visible, with slight differences in peak intensities. The peaks at 1055 cm^{-1} and 773 cm^{-1}, characteristic of the Si–O bond, decrease in strength as the cement replacement percentage increases. This may indicate a reduction in silicate product formation, which affects the mechanical performance of the mortars. This is evident in the mechanical strength results, which is discussed later [9,28,29].

https://doi.org/10.3390/jcs10010042

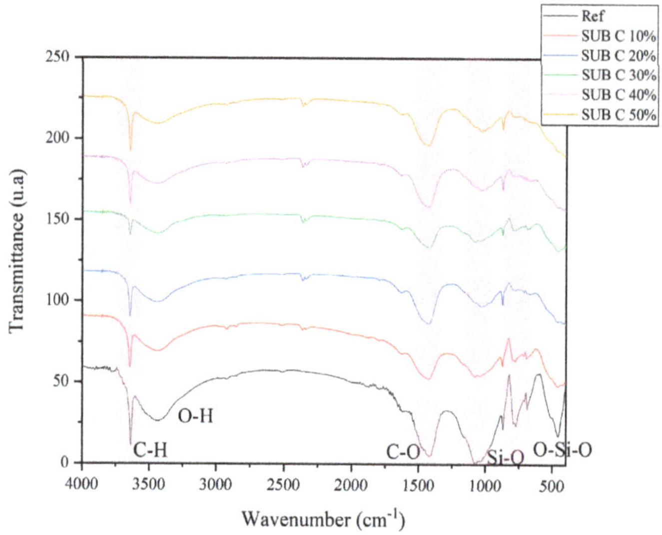

Figure 21. FTIR analysis of reference mortars and mortars with cement replaced by scale.

Finally, Figure 22 presents the results for the partial replacement of sand with scale, showing behavior similar to that of cement replacement, with a noticeable reduction in the intensities of the bands at 1055 cm^{-1} and 773 cm^{-1}, both of which are associated with silica in the system. This decrease is due to the lower content of silicate compounds in the mixture, specifically sand, which can directly influence the microstructure and, consequently, the physical and mechanical properties of the mortars [12,30].

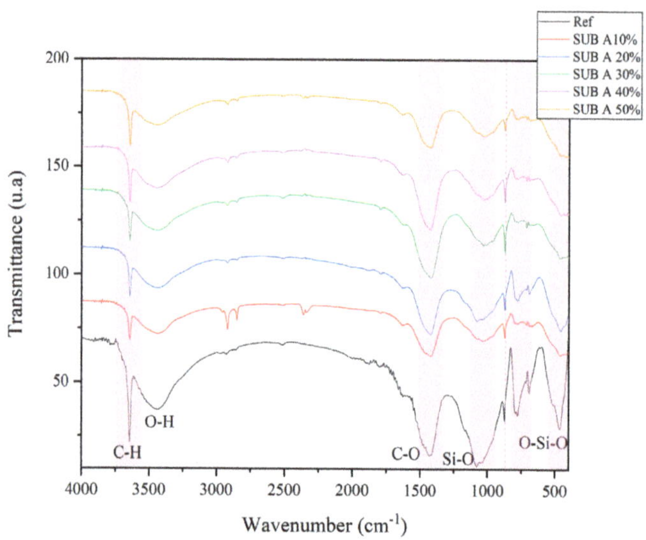

Figure 22. FTIR spectra of the reference mortars and those with sand replaced by scale.

3.5. Microstructural Properties of the Mortars

Figures 23–32 show SEM images of sustainable mortars. When cement is partially replaced with scale, the formation of portlandite, ettringite, and C–S–H remains consistent across all tested proportions. The appearance of portlandite, a compound with low mechan-

https://doi.org/10.3390/jcs10010042

ical strength, is notable. When sand is partially replaced by scale, portlandite, ettringite, and C–S–H are also observed in all proportions.

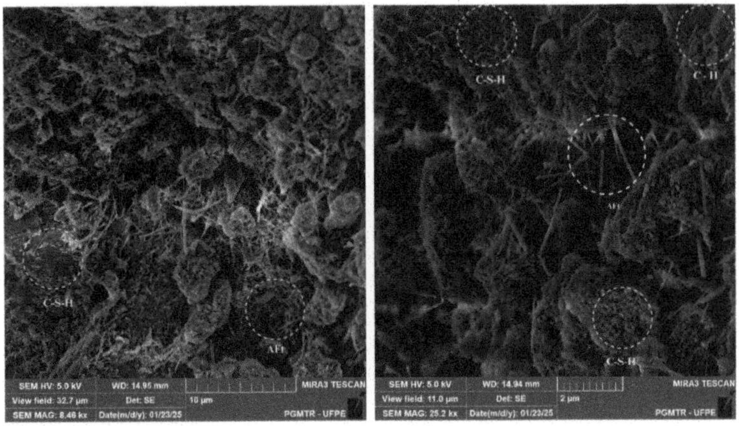

Figure 23. SEM image of the mortar with 10% cement replacement scale.

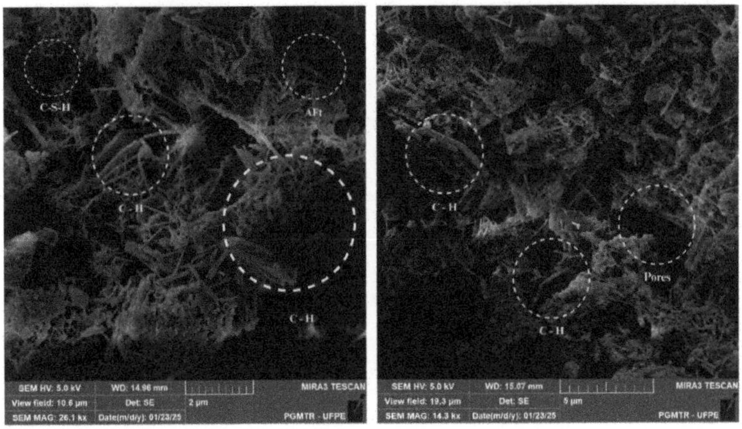

Figure 24. SEM image of the mortar with 20% cement replacement.

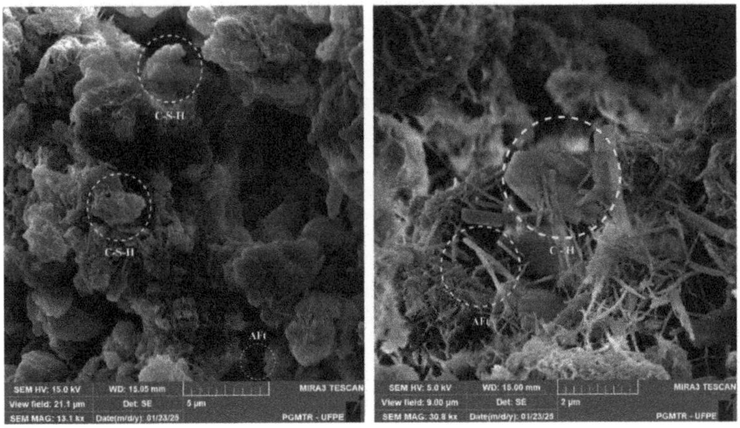

Figure 25. SEM image of the mortar with 30% cement replacement scale.

https://doi.org/10.3390/jcs10010042

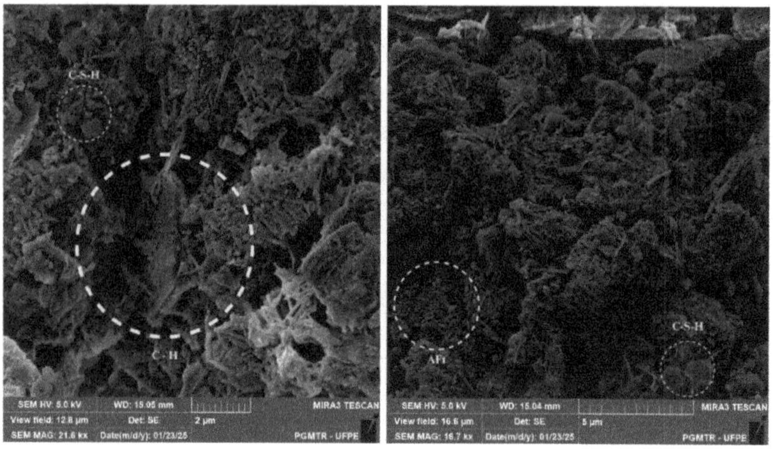

Figure 26. SEM image of the mortar with 40% cement replacement scale.

Figure 27. SEM image of the mortar with 50% cement replacement scale.

Figure 28. SEM image of the mortar with sand replaced by 10% scale.

https://doi.org/10.3390/jcs10010042

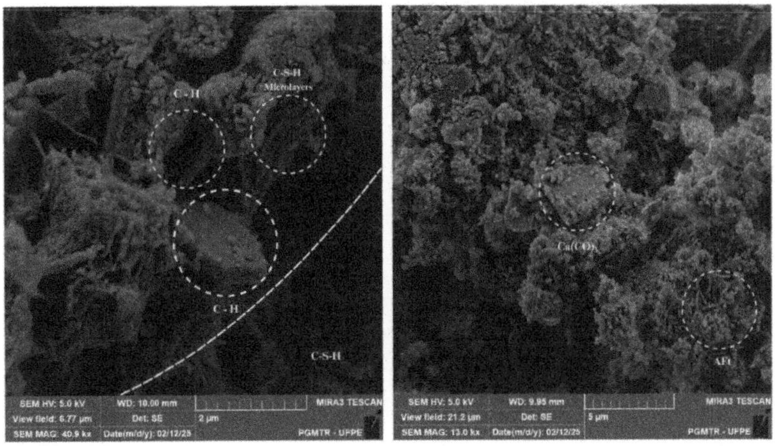

Figure 29. SEM image of the mortar with sand replaced by 20% scale.

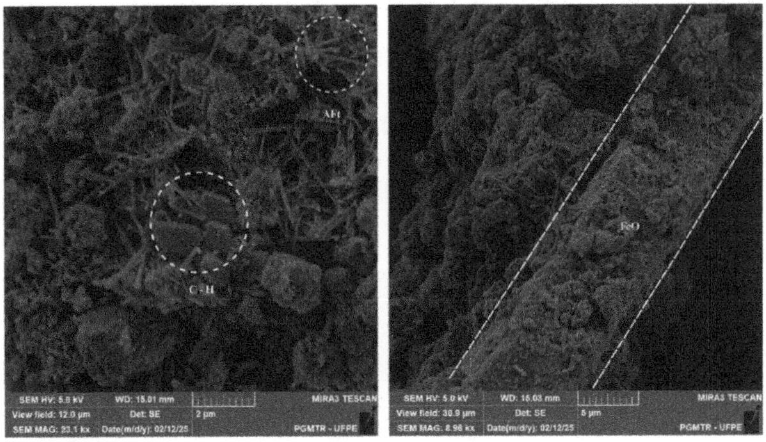

Figure 30. SEM image of the mortar with sand replaced by 30% scale.

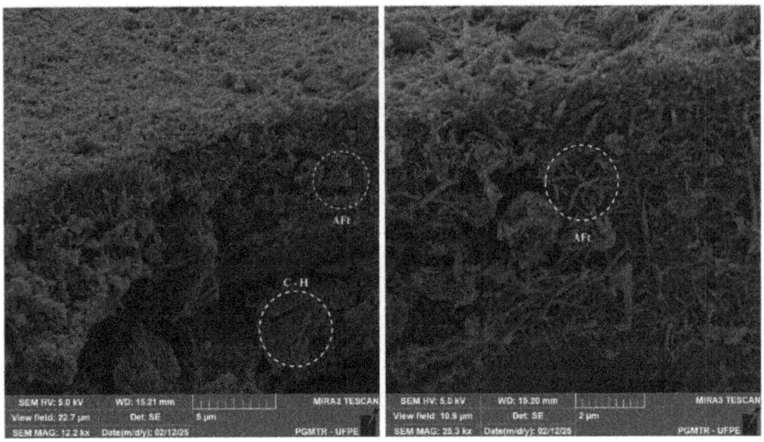

Figure 31. SEM image of the mortar with sand replaced by 40% scale.

https://doi.org/10.3390/jcs10010042

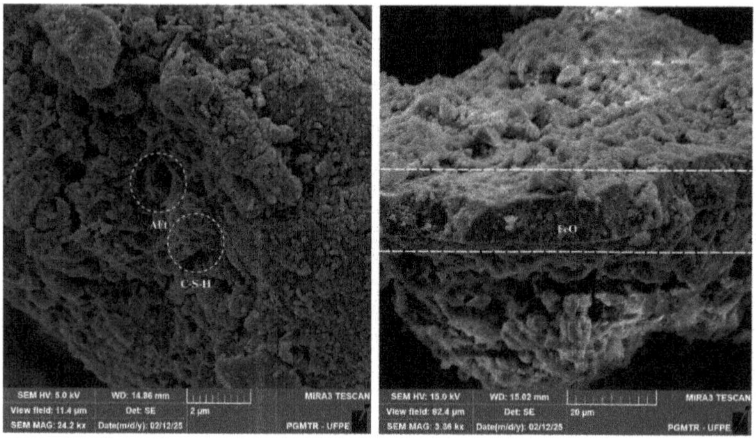

Figure 32. SEM image of the mortar with sand replaced by 50% scale.

Figure 23 (10% Cement Replacement). The micrograph of the mortar with 10% cement replacement demonstrates that the presence of the residue did not inhibit the formation of essential hydrated phases. It is possible to clearly identify the formation of Calcium Silicate Hydrate (C–S–H), along with acicular Ettringite (AFt) crystals and Portlandite (CH) plates. The microstructure suggests that, at this replacement level, the matrix retains reasonable cohesion, as cement dilution is not yet severe enough to prevent the formation of connections among hydration products. However, a reduction in mechanical strength becomes statistically significant.

Figure 24 (20% Cement Replacement). In this image, the persistence of hydration products (C–S–H, CH, and AFt) distributed throughout the matrix is observed. However, the data indicate that intermediate cement replacements (such as 20%) tend to increase capillary porosity and water absorption. The micrograph shows a matrix in which the amount of binder paste decreases proportionally as the inert scale is introduced. The visible presence of Portlandite (CH) is critical, as it is a low-mechanical-strength, high-solubility crystal that contributes to the observed performance reduction relative to the reference.

Figure 25 (30% Cement Replacement). The micrograph for 30% replacement indicates that cement mineral phases still form, but matrix heterogeneity increases. The image highlights C–S–H clusters and Ettringite (AFt) crystals filling spaces. However, since the scale lacks pozzolanic activity (it does not react with Portlandite to form additional C–S–H), it acts merely as an inert filler. The increased proportion of residue "dilutes" the binder, resulting in a less-dense, mechanically weaker microstructure, which aligns with the progressive drop in compressive strength reported in the study.

Figure 26 (40% Cement Replacement). At 40% replacement, the microstructure exhibits a pronounced presence of inert particles within the hydration products. Although C–S–H and Ettringite are still identified, the continuity of the binder phase is interrupted by the high dosage of non-reactive material. Chemical analysis (discussed in the text via XRF and EDS) corroborates that there is a significant reduction in Calcium Oxide (CaO) and an increase in Iron Oxide (Fe_2O_3) in this phase. The image illustrates the "dilution effect": there is less "glue" (hydrated cement) to bind the particles, compromising structural integrity.

Figure 27 (50% Cement Replacement). The image corresponding to the maximum replacement (50%) best illustrates the inert nature of the residue. It is possible to clearly visualize scale particles, identified as Iron Oxides (FeO), adhered to the matrix. The sharp boundary between the FeO and the cement paste, without a diffuse reaction zone, confirms

https://doi.org/10.3390/jcs10010042

the absence of pozzolanic activity. Although the scale is physically incorporated, the drastic reduction in cement (the reactive material) results in a C–S–H-deficient microstructure, which explains why this proportion exhibited the poorest mechanical performance in the study.

In Figures 23 and 27, scale particles adhered to the mortar, marked by iron oxides (FeO). This suggests that the scale acted as an inert material, with no significant pozzolanic activity, but that it may have improved the density and possibly the durability of the mortars [12]. In conclusion, incorporating various types of scale into the essentially inert cementitious matrix can serve as a barrier, thereby promoting pore and microcrack formation.

Furthermore, the spectra obtained indicate that iron (Fe) is more prominently distributed in formulations with higher scale percentages. On the other hand, the silicon (Si) content associated with C–S–H remains relatively constant with addition, with slight reductions when replacements are incorporated. This behavior supports the hypothesis that, in this context, scale does not exert a significant pozzolanic action, serving only as a filler [31,32].

Figure 28 (10% Sand Replacement). The SEM micrograph of the mortar with 10% sand replacement reveals a cohesive microstructural integrity, characterized by the consistent formation of essential hydration products such as Calcium Silicate Hydrate (C–S–H), Portlandite (CH), and Ettringite (AFt). The imagery confirms that at this substitution level, the incorporation of mill scale does not inhibit the hydraulic reactivity of the cement binder, enabling the development of a dense matrix comparable to that of standard formulations. The observed distribution of these phases suggests that the residue acts as a compatible, inert aggregate, maintaining the necessary microstructural cohesion to account for the observed mechanical performance.

Figure 29 (20% Sand Replacement). At the 20% replacement level, the microstructural analysis indicates a denser cementitious matrix, attributed to the higher specific gravity of the mill-scale particles, which promotes tighter solid packing. The micrograph identifies the coexistence of mature hydration phases, including C–S–H microlayers and Calcium Carbonate (Calcite/$CaCO_3$), indicating that the inert filler integrates well without inducing deleterious cracking or void formation. This compact morphology is consistent with physical test results indicating reduced capillary porosity, suggesting that the scale effectively occupies void spaces and enhances the matrix's impermeability.

Figure 30 (30% Sand Replacement). Figure 30 highlights the effectiveness of the Interfacial Transition Zone (ITZ) between the mill scale and the cement paste, demonstrating a "tight physical interface" with no significant detachment voids. The rough surface texture of the scale particles facilitates robust mechanical interlocking with the surrounding C–S–H and Ettringite (AFt) network. This physical adhesion mechanism explains the retention of compressive and tensile strengths observed in the mechanical assays, as the scale functions as a high-stiffness inclusion that, when properly encapsulated, reinforces the composite structure rather than acting as a defect.

Figure 31 (40% Sand Replacement). At 40% substitution, the micrograph clearly delineates the distinct boundary between the inert phase and the binder, accentuating scale particles rich in Iron Oxides (FeO) embedded within the C–S–H matrix. This observation confirms the non-pozzolanic nature of the residue; it does not chemically react to form new binding phases but instead serves as a dense, impermeable physical barrier. The presence of these laminar, metallic-mineral particles disrupts the continuity of the capillary pore network, thereby increasing tortuosity and impeding fluid transport, which directly correlates with enhanced durability in water absorption.

Figure 32 (50% Sand Replacement). The analysis of the mortar with 50% replacement reveals a microstructure dominated by a high volume of dense mill-scale particles (FeO)

https://doi.org/10.3390/jcs10010042

that appear as compact blocks. Despite the substantial load of inert material, the image demonstrates that the cement paste maintains adequate coverage of the iron particles, with C–S–H and Ettringite (AFt) detectable at the interfaces. While the high fines content may present rheological challenges, the micrograph indicates that the material's high density (5.5 g/cm^3) yields a low-permeability composite, in which the residue acts as a high-density filler that physically blocks the ingress of aggressive agents.

The SEM/EDS data for mortars with cement replacement (Figures 32–39) show that the percentages of calcium (Ca), oxygen (O), carbon (C), and silicon (Si) display expected patterns that repeat across different proportions. Regarding sand replacement, due to operational limitations, EDS was obtained only for 10%, 20%, and 40% sand replacement. These results indicated a significant increase in oxygen (O) and iron (Fe) concentrations, since the sand was replaced in substantial quantities [33–35].

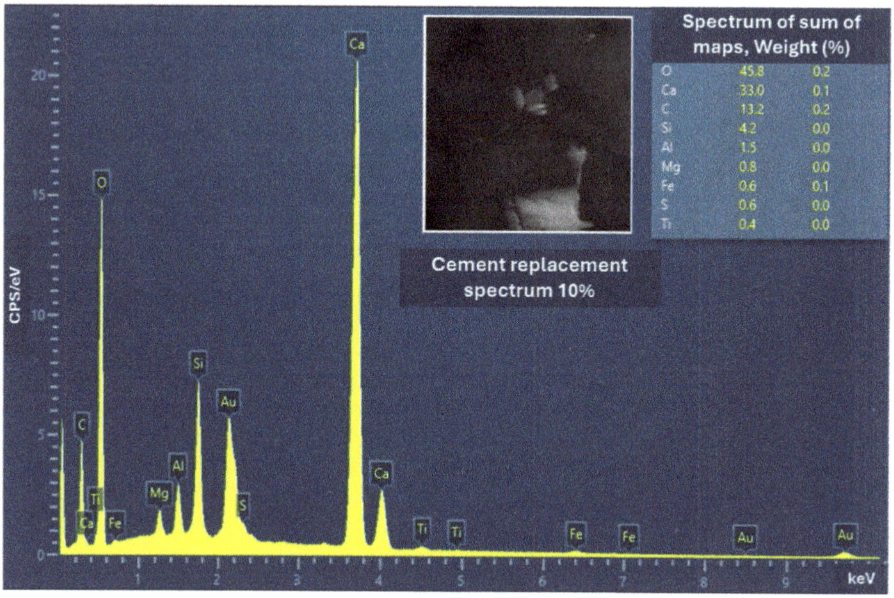

Figure 33. EDS image of the mortar with 10% cement replacement by scale residue.

Figure 33 (EDS of 10% Cement Replacement). The EDS spectrum for the 10% cement replacement exhibits a chemical profile dominated by Calcium (Ca, 33.0%) and Oxygen (O, 45.8%), which is characteristic of a standard cementitious matrix rich in Calcium Silicate Hydrate (C–S–H) and Portlandite1. The Iron (Fe) content remains low at 0.6%, indicating that at this low replacement level, the mill scale is well dispersed and does not significantly alter the elemental balance of the hydrated paste. The strong Calcium peak indicates that the binder's chemistry is preserved.

Figure 34 (EDS of 20% Cement Replacement). At 20% replacement, the spectrum remains chemically similar to the 10% formulation and the reference, with Calcium (Ca) rising to 38.5% and Silicon (Si) at 3.0%. The Iron (Fe) content remains stable at 0.6%, suggesting that the scale particles are effectively encapsulated within the matrix. The persistence of high Calcium peaks corroborates the SEM findings that hydration products are still the primary constituents, despite the initial dilution of the cement.

https://doi.org/10.3390/jcs10010042

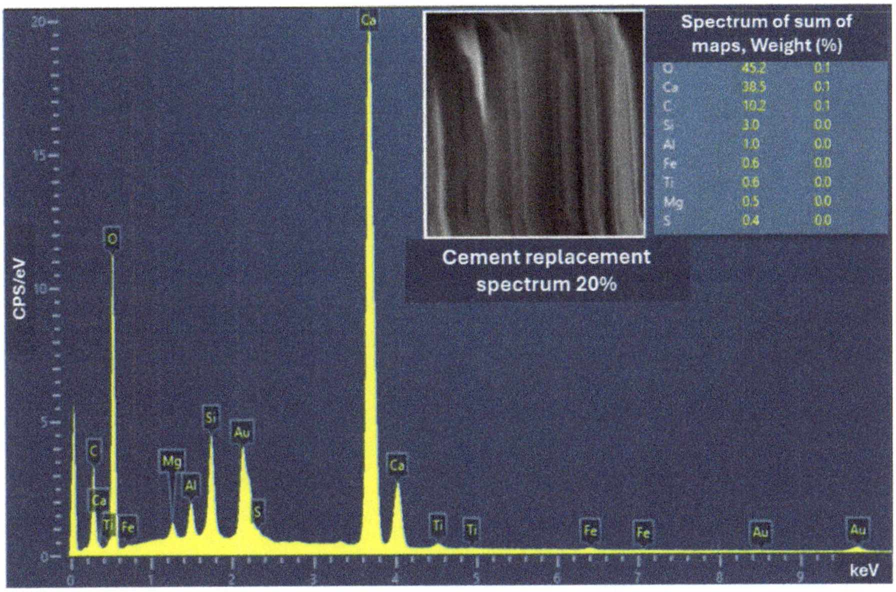

Figure 34. EDS image of the mortar with 20% cement replacement by scale residue.

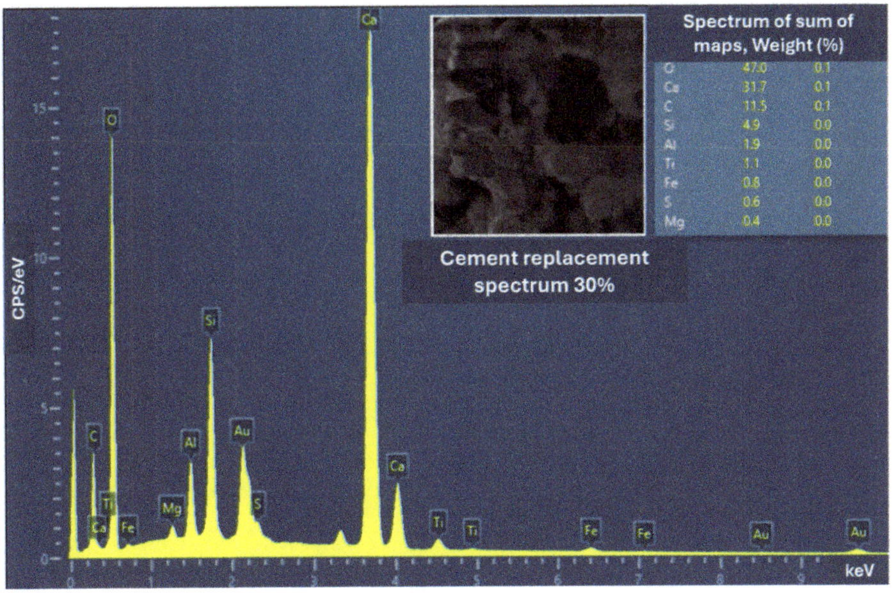

Figure 35. EDS image of the mortar with 30% cement replacement by scale residue.

Figure 35 (EDS of 30% Cement Replacement). The analysis of the 30% replacement sample indicates a slight shift in elemental composition, with Iron (Fe) increasing marginally to 0.8%. While the Calcium (Ca, 31.7%) and Oxygen (O, 47.0%) levels continue to indicate the presence of hydration phases, the slight rise in iron signals the increasing presence of the residue6. This gradual chemical change is supported by the mechanical data, in which binder dilution begins to affect the matrix's overall cohesion.

https://doi.org/10.3390/jcs10010042

Figure 36 (EDS of 40% Cement Replacement). In this spectrum, the Iron (Fe) content doubles to 1.2% compared to previous levels, while Calcium (Ca) remains dominant at 41.2%. This increase reflects the higher volume of mill scale replacing the cement. The detection of distinct iron signals provides evidence for the physical presence of iron oxide particles, which disrupt the homogeneity of the calcium–silicate binder and are consistent with the "dilution effect" discussed in the text regarding mechanical strength loss.

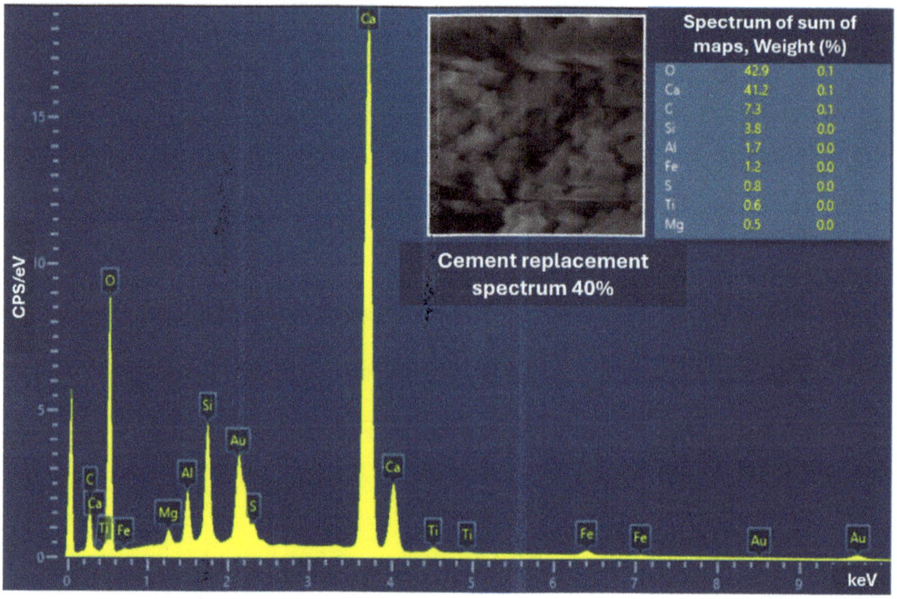

Figure 36. EDS image of the mortar with 40% cement replacement by scale residue.

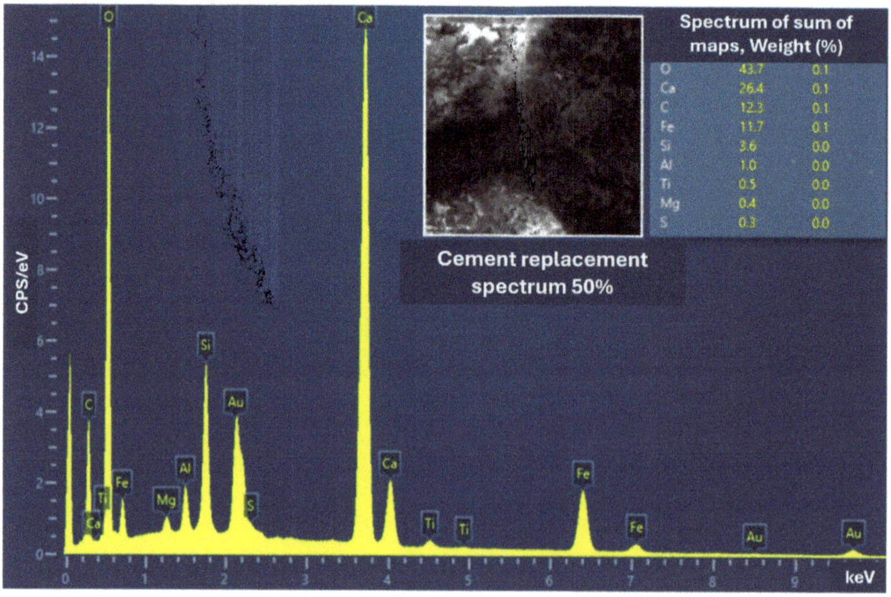

Figure 37. EDS image of the mortar with 50% cement replacement by scale residue.

https://doi.org/10.3390/jcs10010042

Figure 37 (EDS of 50% Cement Replacement). The spectrum for 50% replacement indicates a marked change in the chemical profile, with the Iron (Fe) content increasing to 11.7%. This sharp increase confirms the massive presence of iron oxides (Fe_2O_3, FeO) substituting the hydraulic binder. The spectral shape changes to reflect this high impurity load, indicating that the matrix is now predominantly composed of inert filler rather than reactive cement, which directly explains the observed minimum compressive strength.

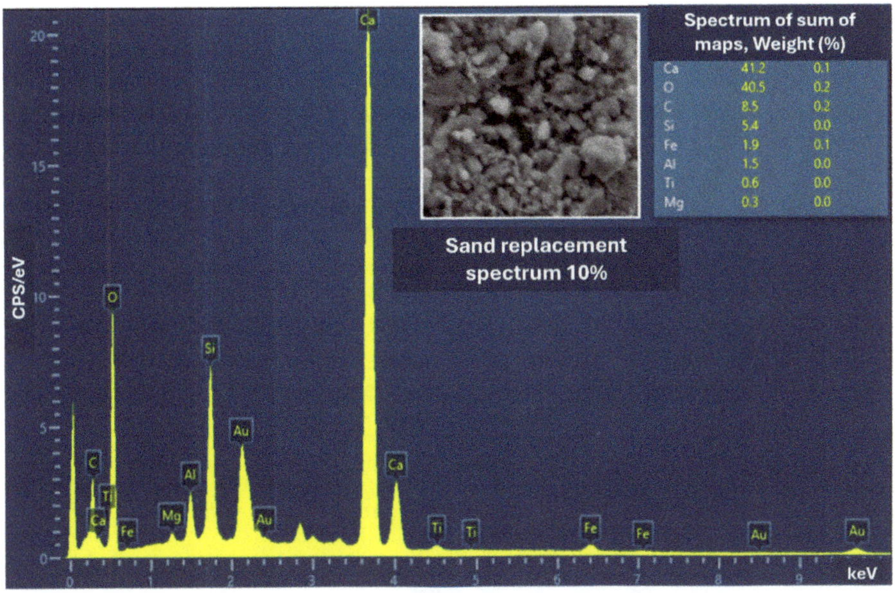

Figure 38. Image of the EDS analysis of the mortar with 10% sand replacement by scale.

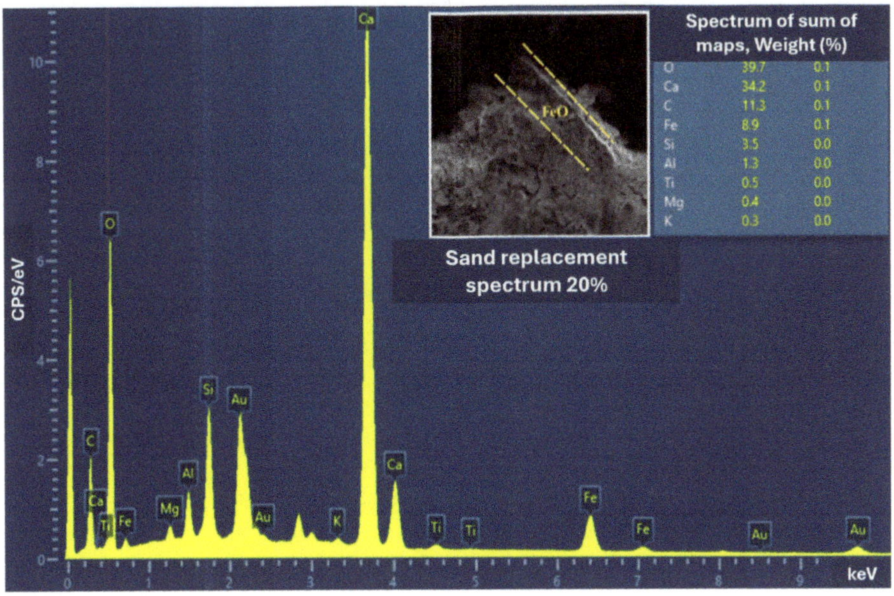

Figure 39. Image of the EDS analysis of the mortar with 20% sand replacement by scale.

https://doi.org/10.3390/jcs10010042

Figure 38 (EDS of 10% Sand Replacement). Switching to the sand replacement route, the 10% sample shows a higher Iron (Fe) content (1.9%) than the equivalent cement replacement. This is expected because the residue is replacing silica (SiO_2) rather than the binder. The spectrum shows a balanced distribution of Calcium (Ca, 41.2%) from the cement paste and Silicon (Si, 5.4%) from the remaining sand, indicating a healthy matrix in which the scale acts as a supplementary fine aggregate.

Figure 39 (EDS of 20% Sand Replacement). The 20% sand replacement spectrum reveals a substantial increase in Iron (Fe) to 8.9%. The inset mapping is particularly revealing, showing a distinct iron-rich region (marked as FeO) adjacent to the calcium-rich paste. This confirms the formation of a tight physical interface without chemical diffusion, validating the "inert filler" mechanism. The Calcium (Ca) level (34.2%) indicates that the cement paste remains chemically intact around the scale particles, thereby maintaining its binding capacity.

Figure 40 (EDS of 40% Sand Replacement). The final spectrum for 40% sand replacement shows a large Iron (Fe) peak at 31.0%, consistent with the significant substitution of natural sand by metallic waste. The Oxygen (O) level (32.4%), combined with the high iron content, confirms the predominance of iron oxides (Fe_2O_3, Fe_3O_4). Crucially, despite this high iron load, the Calcium (Ca) signal (15.0%) indicates the continued presence of C–S–H gel binding the aggregates. This supports the conclusion that sand replacement is the technically superior approach, as it permits high waste incorporation without chemically depleting the binder.

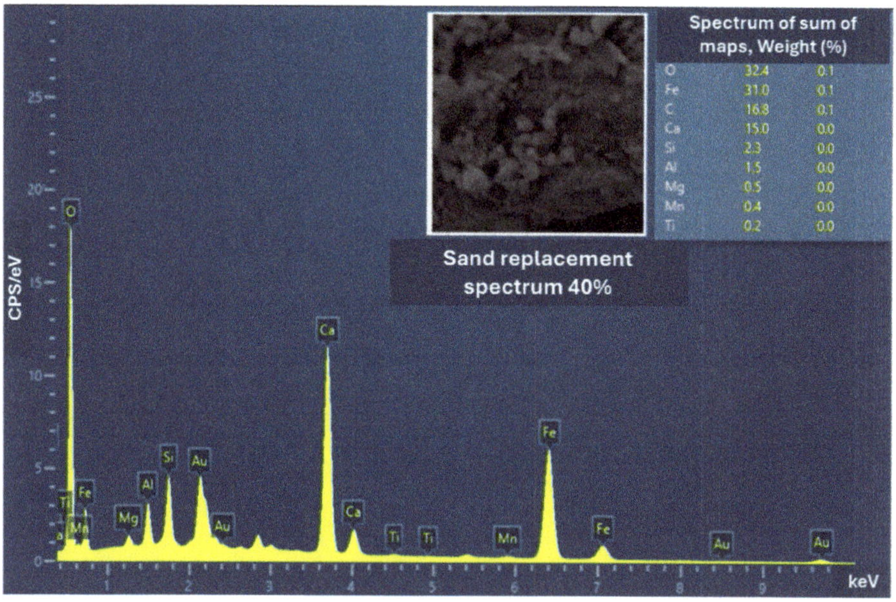

Figure 40. Image of the EDS analysis of the mortar with 40% sand replacement by scale.

Energy-dispersive spectroscopy (EDS) of sand showed a steady increase in oxygen (O) and iron (Fe) levels in samples with 10%, 20%, and 40% scale replacement. This reflects the increasing incorporation of iron oxides (Fe_2O_3, Fe_2O_4, and Fe_2O_5), which are the primary components of this steelmaking process, as widely documented in the literature. Several authors describe the scale as predominantly ferric and high in iron, which explains the Fe levels of 1.9%, 8.9%, and 31.0% at the respective replacement amounts. These findings confirm theoretical predictions and support the observed link between scale content and

https://doi.org/10.3390/jcs10010042

iron enrichment in the cementitious matrix, since these particles remain as inert oxides and do not significantly participate in hydration reactions [36,37].

From a microstructural perspective, incorporating scale tends to modify the interfacial transition zone (ITZ) and matrix density due to the higher specific gravity of the replacement material and its rough surface, which can promote paste-aggregate adhesion. Studies also report that moderate substitutions can favor porosity reduction and the production of denser, less permeable matrices, thereby enhancing durability mechanisms. In contrast, high levels may necessitate rheological and consolidation adjustments, potentially resulting in a loss of mechanical performance. Thus, the EDS results are consistent with the behavior described in the literature, reinforcing that scale acts primarily as a semi-inert mineral phase, altering the physical and microstructural properties of the matrix without modifying the essential cement hydration mechanisms [9,33].

Thus, the results suggest that the technical feasibility of scaling depends on its use as a partial replacement for fine aggregate, particularly in the 10–40% range, where microstructural improvements yield better performance or, at a minimum, results comparable to those of reference [26]. At the same time, the cement replacement route should be avoided for structural applications or for those requiring higher mechanical stresses, unless combined with active additives that mitigate reactivity loss, such as fine pozzolans or silica fume. From a sustainability standpoint, sand replacement is the most effective option among the evaluated routes due to its enhanced performance, reduced natural resource extraction, and alignment with circular economy principles [10,38,39].

A closer examination of the EDS elemental mapping reveals a distinct interaction at the Interfacial Transition Zone (ITZ). The sharp contrast between the iron-rich regions (red areas), representing mill-scale particles, and the calcium-rich regions (pink areas), corresponding to the hydrated cement paste, confirms the predominantly inert nature of the waste [15]. The absence of a diffuse transition zone suggests limited chemical reaction between the scale and the cement matrix. However, the microstructure indicates a tight physical interface without significant detachment voids. This observation supports the mechanical behavior results: while the lack of chemical reactivity explains the strength reduction in cement replacement (where binding capacity is lost), the effective physical interlocking observed at the ITZ explains why sand replacement formulations maintained their mechanical performance [9,39].

The correlation between the microstructural analysis and the physical–mechanical tests provides a clear explanation for the divergent behaviors observed in the two substitution routes:

- The Mechanism of Strength Retention in Sand Replacement (SUB-A): As evidenced by the SEM micrographs (Figures 28–32) and EDS mapping (Figures 38–40), the mill scale particles exhibit a rough texture and a high degree of compactness within the matrix. Although XRD (Figure 20) and FTIR (Figure 22) confirm the inert chemical nature of the scale (with no new hydration products formed), the interface between the scale and the cement paste is tight and free of significant detachment voids. This effective physical interlocking at the Interfacial Transition Zone (ITZ) explains why the compressive and tensile strengths were maintained (Figures 14 and 16) despite the replacement of natural sand. The scale acts as a high-stiffness inclusion that mechanically reinforces the matrix, provided it is well-encapsulated.
- The Mechanism of Permeability Reduction: The capillary absorption results (Figure 12), which showed a decrease in permeability for the SUB-A group, are directly explained by the EDS findings. The iron-rich scale particles are denser (5.5 g/cm^3) and less porous than the natural sand they replaced. In the microstructural images, these particles appear as dense blocks that interrupt the continuity of capillary pores. Con-

https://doi.org/10.3390/jcs10010042

sequently, they act as impermeable physical barriers, increasing the tortuosity of the pore network and hindering water transport, both of which are critical to the enhanced durability potential of these composites.

4. Conclusions

The results demonstrated that the use of industrial lamination waste (lamination scale) in mortars is a technical and environmentally viable alternative, aligning with the principles of sustainable development and circular economy in civil construction. The technical feasibility, however, depends directly on the method used to incorporate waste. The study demonstrated that the scale primarily functions as an inert filler, exhibiting no significant pozzolanic activity. Microstructural (SEM, XRD, FTIR) and chemical (XRF, EDS) analyses confirmed this nature.

The main scientific findings were:

- Cement Replacement (SUB C): This approach compromised the performance of the mortar.
    - Mechanical Properties: There was a progressive decrease in axial compressive strength and tensile strength at flexure as the replacement percentage increased. XRD and FTIR analyses revealed reduced formation of hydration products, as evidenced by decreases in the portlandite (P) and alite (C_3S) peaks and in Si–O bonding bands. Chemically, a reduction in CaO content and an increase in Fe_2O_3 content were observed.
    - Physical Properties: Although workability (consistency index) was maintained or slightly increased, water absorption by capillary increased in intermediate proportions (20%, 30% and 40%) compared to references.
    - Conclusion (SUB C): The cement replacement route should be avoided for applications that require mechanical performance.
- Sand Substitution (SUB A): This approach offered the best balance for hardened state properties but presented a clear trade-off regarding fresh state performance.
    - Mechanical and Physical Balance: Unlike cement replacement, substituting sand did not compromise the hydration process, maintaining compressive and flexural strength levels comparable to the reference up to 40% replacement. Furthermore, due to the high density of mill scale (5.5 g/cm^3), the matrix became denser and less permeable, with 20% and 30% proportions exhibiting the best durability indicators (i.e., the lowest capillary absorption).
    - Practical Limitations: However, a significant reduction in workability (consistency index) was observed as the scale content increased. This drastic drop represents a practical challenge for on-site applications.

Final Considerations: Consequently, the use of mill scale as a fine aggregate is technically feasible and environmentally beneficial, particularly within the 10–40% range. However, this recommendation is contingent on effective management of rheological properties. For high replacement levels, the use of water-reducing admixtures (plasticizers) is strongly recommended to recover workability without increasing the water-to-binder ratio, thereby preserving the observed mechanical gains.

Study Limitations and Future Research: It is important to note that this study focused on the physical–mechanical feasibility over 28 days. Although the reduced capillary absorption observed suggests potential for high durability (by limiting the ingress of water and oxygen), specific long-term behaviors—such as the risk of iron oxide leaching (rust staining), volume stability under thermal cycles, and resistance to carbonation—were not evaluated. Therefore, future research must conduct accelerated aging and durability tests

https://doi.org/10.3390/jcs10010042

to fully validate the service life of these sustainable composites before their widespread structural application.

Author Contributions: Conceptualization, A.L.M.A. and N.B.D.L.; methodology, A.L.M.A., J.V.B.L.O., R.C.M.C., B.S.T., E.C.C., N.B.L., K.G.B.A. and N.B.D.L., formal analysis, A.L.M.A., J.V.B.L.O., R.C.M.C., B.S.T., E.C.C., N.B.L., K.G.B.A. and N.B.D.L., investigation, A.L.M.A. and N.B.D.L., resources, N.B.D.L.; data curation, A.L.M.A., J.V.B.L.O., R.C.M.C., B.S.T., E.C.C., N.B.L., K.G.B.A. and N.B.D.L., writing—original draft preparation, A.L.M.A., J.V.B.L.O., R.C.M.C., B.S.T., E.C.C., N.B.L., K.G.B.A. and N.B.D.L. All authors have read and agreed to the published version of the manuscript.

Funding: This research was funded by CNPq (funding number 310132/2021-5), CAPES (88887.939696/2024-00), and FACEPE (funding number APQ-0079-3.01/24).

Data Availability Statement: The original contributions presented in this study are included in the article. Further inquiries can be directed to the corresponding author.

Acknowledgments: The authors are grateful for the financial support of the Brazilian Agencies CNPq, CAPES, and FACEPE. N.B.D.L. thanks the L'Oréal-UNESCO-ABC "For Women in Science".

Conflicts of Interest: The authors declare no conflicts of interest.

Abbreviations

The following abbreviations are used in this manuscript:

ABNT	Associação Brasileira de Normas Técnicas
XRF	X-ray fluorescence
XRD	X-ray diffraction
EDS	energy-dispersive spectroscopy
SEM	Scanning Electron Microscopy
ITZ	Interfacial transition zone
SAW	Submerged arc welding
FTIR	Fourier-transform infrared spectroscopy

References

1. Baptista Junior, J.V.; Romanel, C. Sustentabilidade na indústria da construção: Uma logística para reciclagem dos resíduos de pequenas obras. *Urbe-Revista Bras. Gestao Urbana* **2013**, *5*, 27. [CrossRef]
2. Pagio, M.Z.; Carrareto, L.F.; Vieira, G.L.; Magalhães, D.C. Caracterização de resíduos siderúrgicos visando à aplicação em matrizes cimentícias. *Ambiente Construído* **2022**, *22*, 167–186. [CrossRef]
3. He, Y.; Jia, S.; Yi, H.; Tang, X.; Yu, Q.; Gao, F.; Kang, D.; Zhao, S. Utilization of steel slag in air pollution and greenhouse gas emission reduction-application, mechanism and challenge: A review. *J. Environ. Chem. Eng.* **2024**, *12*, 5. [CrossRef]
4. Kong, F.; Ying, Y.; Lu, S. Heavy metal pollution risk of desulfurized steel slag as a soil amendment in cycling use of solid wastes. *J. Environ. Sci.* **2023**, *127*, 349–360. [CrossRef]
5. Moraes, C.A.M.; Evaldt, D.C.; Camacho, A.L.D.; Kich, A.M. Caracterização de Carepa Metálica Visando Sua Valorização Como Coproduto. In Proceedings of the Congresso Anual Da Abm, Belo Horizonte, Brazil, 30 July 2013; ABM: São Paulo, Brazil, 2013. [CrossRef]
6. Gagliotti, A.B. Análise da Reciclagem de Carepa de Aço Por Meio de Processos de Redução. Bachelor's Thesis, Universidade Tecnológica Federal do Paraná, Londrina, Brazil, 2019.
7. Caus, G.D.; Bertolini, G.R.F.; Bulhões, R. Diagnóstico da gestão de resíduos sólidos: Estudo de caso em uma indústria metalúrgica. *Rev. Micro Pequena Empresa São Paulo* **2024**, *18*, 178–196. [CrossRef]
8. Pereira, F.M.; de Verney, J.C.K.; Lenz, D.M. Avaliação do emprego de carepa de aço como agregado miúdo em concreto. *REM Rev. Esc. Minas* **2011**, *64*, 463–469. [CrossRef]
9. Siriwattanakarn, A.; Wongsa, A.; Eua-Anant, N.; Sata, V.; Sukontasukkul, P.; Chindaprasirt, P. Utilization of Mill Scale Waste as Natural Fine Aggregate Replacement in Mortar: Evaluation of Physical, Mechanical, Durability, and Post-Fire Properties. *Recycling* **2025**, *10*, 20. [CrossRef]
10. Kushwah, R.; Gangotiya, S.; Patil, K. A Review on Engineering Behaviour of Sustainable Concrete with Steel Mill Scale and the Effect of Admixture on Workability Improvement. *Int. J. Res. Appl. Sci. Eng. Technol.* **2024**, *12*, 859–866. [CrossRef]

https://doi.org/10.3390/jcs10010042

11. Ozturk, M.; Depci, T.; Bahceci, E.; Karaaslan, M.; Akgol, O.; Sevim, U.K. Production of new electromagnetic wave shielder mortar using waste mill scales. *Constr. Build. Mater.* **2020**, *242*, 118028. [CrossRef]
12. Parvathikumar, G.; Sahadevan, B.; Palanisamy, C. Scrap steel mill scale as river sand replacement in cement concrete: Effect on durability characteristics. *J. Mater. Cycles Waste Manag.* **2024**, *26*, 1490–1504. [CrossRef]
13. Ahmed, Y.; Hessien, M.; Rashad, M.; Ibrahim, I. Nano-crystalline copper ferrites from secondary iron oxide (mill scale). *J. Magn. Magn. Mater.* **2009**, *321*, 181–187. [CrossRef]
14. Liu, H.; Lin, H.; Liu, X.; Wang, J.; Pang, X.; Cui, S.; Kong, X. Mechanism of accelerated self-healing behavior of cement mortars incorporating triethanolamine: Carbonation of portlandite. *Constr. Build. Mater.* **2021**, *308*, 125050. [CrossRef]
15. *NBR 17054*; Agregados—Determinação da Composição Granulométrica—Método de Ensaio. Associação Brasileira de Normas Técnicas: Rio de Janeiro, Brazil, 2022.
16. *NBR 7215*; Portland Cement: Determination of Compressive Strength. Associação Brasileira de Normas Técnicas: Rio de Janeiro, Brazil, 2019.
17. *NBR 13279*; Mortar for Laying and Covering Walls and Ceilings—Determination of Tensile Strength in Flexure and Compression. Associação Brasileira de Normas Técnicas: Rio de Janeiro, Brazil, 2005.
18. *NBR 13278*; Mortar for Laying and Covering Walls and Ceilings—Determination of Mass Density and Incorporated Air Content. Associação Brasileira de Normas Técnicas: Rio de Janeiro, Brazil, 2005.
19. *NBR 9779*; Hardened Mortar and Concrete—Determination of Water Absorption by Capillarity. Associação Brasileira de Normas Técnicas: Rio de Janeiro, Brazil, 2012.
20. *NBR 7222*; Concrete and Mortar—Determination of Tensile Strength by Diametrical Compression of Cylindrical Specimens. Associação Brasileira de Normas Técnicas: Rio de Janeiro, Brazil, 2011.
21. *NBR 13276*; Mortar for Laying and Covering Walls and Ceilings—Determination of the Consistency Index. Associação Brasileira de Normas Técnicas: Rio de Janeiro, Brazil, 2016.
22. Girskas, G.; Kligys, M. Research on the main properties of cementitious mortars prepared with high-Fe_2O_3-content raw drinking water treatment sludge. *Materials* **2025**, *18*, 759. [CrossRef] [PubMed]
23. Omrane, M.; Rabehi, M. Effect of natural pozzolan and recycled concrete aggregates on thermal and physico-mechanical characteristics of self-compacting concrete. *Constr. Build. Mater.* **2020**, *247*, 118576. [CrossRef]
24. Mehta, P.K.; Monteiro, P.J.M. *Concreto: Microestrutura, Propriedades e Materiais*, 2nd ed.; IBRACON: São Paulo, Brazil, 2014; p. 751, ISBN 978-85-98576-213-3.
25. Taylor, H.F.W. *Cement Chemistry*, 2nd ed.; Thomas Telford: London, UK, 1997.
26. Veloso, C.K.d.S.; Pertile, B.H.P.; Nascimento, L.G.D.; Soares, T.K.F.; Lopes, P.D.; Moura, A.O.C.; Sousa, R.M.L.; Melo, S.T. Avaliação do desempenho da argamassa com incorporação de resíduos de materiais cerâmicos. *Braz. J. Dev.* **2020**, *6*, 822–837. [CrossRef]
27. Ghrair, A.M.; Said, A.J.; Aldaoud, N.; Miqdadi, R.; Ahmad, A.A.L. Characterisation and Reverse Engineering of Eco-friendly Historical Mortar: Qasr Tuba, Desert Castles in Jordan. *J. Ecol. Eng.* **2021**, *22*, 121–134. [CrossRef]
28. Ghazali, N.; Muthusamy, K.; Embong, R.; Rahim, I.S.A.; Razali, N.F.M.; Yahaya, F.M.; Ariffin, N.F.; Ahmad, S.W. Effect of Fly Ash as Partial Cement Replacement on Workability and Compressive Strength of Palm Oil Clinker Lightweight Concrete. *IOP Conf. Ser. Earth Environ. Sci.* **2020**, *682*, 012038. [CrossRef]
29. Akula, P.; Little, D.N. Mineralogical characterization and thermodynamic modeling of synthesized ettringite from Ca-Al-SO_4 suspensions. *Constr. Build. Mater.* **2021**, *269*, 121304. [CrossRef]
30. Neville, A.M. *Propriedades do Concreto*, 5th ed.; Bookman: Porto Alegre, Brazil, 2016.
31. Pandini, N.T. Influência da relação Mássica CaO/SiO_2 Nas Propriedades Tecnológicas de Argamassas à Base de Cinza Volante Álcali-Ativada Com Cal Virgem. Master's Thesis, Universidad e Estadual do Norte Fluminense Darcy Ribeiro, Campos dos Goytacazes, Brazil, 2019.
32. Boudache, S.; Loukili, A.; Izoret, L.; Rozière, E. Investigating the role played by portlandite and C-A-S-H in the degradation response of pozzolanic and slag cements to external sulphate attack. *J. Build. Eng.* **2023**, *67*, 106053. [CrossRef]
33. Mohr, B.J.; Islam, M.S.; Bryant, L. Long-term behavior of mortars experiencing delayed ettringite formation. *Cement* **2024**, *16*, 100104. [CrossRef]
34. Wang, J.; Gao, C.; Tang, J.; Hu, Z.; Liu, J. The multi-scale mechanical properties of calcium-silicate-hydrate. *Cem. Concr. Compos.* **2023**, *140*, 105097. [CrossRef]
35. Arnold, M.C.; de Vargas, A.S.; Bianchini, L. Study of electric-arc furnace dust (EAFD) in fly ash and rice husk ash-based geopolymers. *Adv. Powder Technol.* **2017**, *28*, 2023–2034. [CrossRef]
36. Kupwade-Patil, K.; Palkovic, S.D.; Bumajdad, A.; Soriano, C.; Büyüköztürk, O. Use of silica fume and natural volcanic ash as a replacement to Portland cement: Micro and pore structural investigation using NMR, XRD, FTIR and X-ray microtomography. *Constr. Build. Mater.* **2018**, *158*, 574–590. [CrossRef]
37. Teti, B.; Amorim, A.; Costa, E.; Alves, K.; Lima, N. Incorporating industrial residue of submerged arc welding (SAW) in cement-based mortar matrices as a green strategy. *Next Sustain.* **2025**, *5*, 100080. [CrossRef]

https://doi.org/10.3390/jcs10010042

38. Scrivener, K.L. Backscattered electron imaging of cementitious microstructures: Understanding and quantification. *Cement Concrete Compos.* **2004**, *26*, 935–945. [CrossRef]
39. You, N.; Chen, Z.; Gao, Z.; Song, X. The effect of copper slag as a precursor on the mechanical properties, shrinkage and pore structure of alkali-activated slag-copper slag mortar. *J. Build. Eng.* **2024**, *98*, 111151. [CrossRef]

Disclaimer/Publisher's Note: The statements, opinions and data contained in all publications are solely those of the individual author(s) and contributor(s) and not of MDPI and/or the editor(s). MDPI and/or the editor(s) disclaim responsibility for any injury to people or property resulting from any ideas, methods, instructions or products referred to in the content.

https://doi.org/10.3390/jcs10010042

Journal of
Composites Science

MDPI

Correction

Correction: Feng et al. Dynamic Compressive Behavior of CFRP-Confined High Water Material. *J. Compos. Sci.* 2025, *9*, 482

Feiyang Feng [1], Shuling Meng [2], Haishan Huang [3], Yafei Zhou [4] and Hongchao Zhao [5,*]

1 School of Geology and Mining Engineering, Xinjiang University, Urumqi 830047, China; 107552404778@stu.xju.edu.cn
2 China West Construction Group Co., Ltd., Urumqi 610200, China; 15037445129@163.com
3 Xinjiang Saier Energy & Mining Co., Ltd., Tacheng 834406, China; 18999317182@163.com
4 Korla Jinchuan Mining Co., Ltd., Korla 841001, China; zhouyafei25400@163.com
5 State Key Laboratory of Intelligent Construction and Healthy Operation and Maintenance of Deep Underground Engineering, Xuzhou 221116, China
* Correspondence: zhaohongchao@xju.edu.cn

Figure

In the original publication [1], there was a mistake in **Figure 5. Strain rate–stress–energy absorption curves under different water–cement ratios (The line graph indicates peak stress, while the bar chart represents absorbed energy.).** The mistake is Figure 5, which is a duplicate of Figure 4 due to an oversight. The corrected **Figure 5** appears below. The authors state that the scientific conclusions are unaffected. This correction was approved by the Academic Editor. The original publication has also been updated.

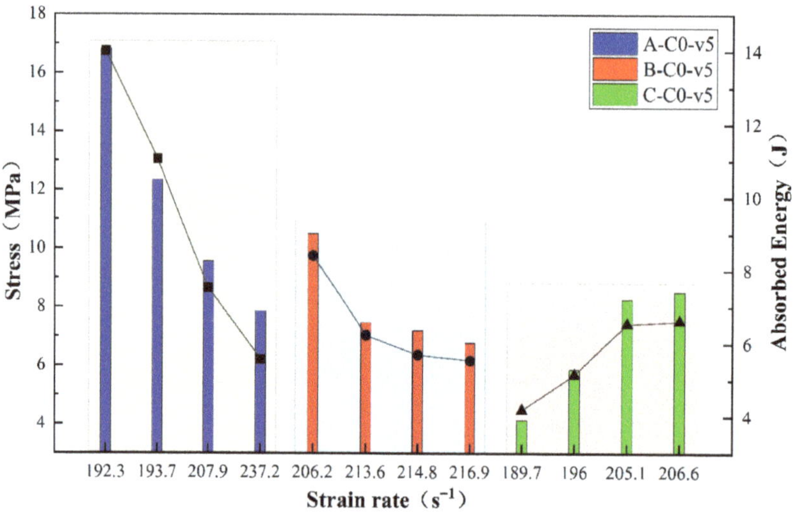

Figure 5. Strain rate–stress–energy absorption curves under different water–cement ratios (The line graph indicates peak stress, while the bar chart represents absorbed energy.).

Reference

1. Feng, F.; Meng, S.; Huang, H.; Zhou, Y.; Zhao, H. Dynamic Compressive Behavior of CFRP-Confined High Water Material. *J. Compos. Sci.* **2025**, *9*, 482. [CrossRef]

Disclaimer/Publisher's Note: The statements, opinions and data contained in all publications are solely those of the individual author(s) and contributor(s) and not of MDPI and/or the editor(s). MDPI and/or the editor(s) disclaim responsibility for any injury to people or property resulting from any ideas, methods, instructions or products referred to in the content.

MDPI AG
Grosspeteranlage 5
4052 Basel
Switzerland
Tel.: +41 61 683 77 34

Journal of Composites Science Editorial Office
E-mail: jcs@mdpi.com
www.mdpi.com/journal/jcs

Disclaimer/Publisher's Note: The title and front matter of this reprint are at the discretion of the Guest Editors. The publisher is not responsible for their content or any associated concerns. The statements, opinions and data contained in all individual articles are solely those of the individual Editors and contributors and not of MDPI. MDPI disclaims responsibility for any injury to people or property resulting from any ideas, methods, instructions or products referred to in the content.

www.ingramcontent.com/pod-product-compliance
Lightning Source LLC
Chambersburg PA
CBHW061513210926
51850CB00008B/1908
9783725871629